INTERSECT:
A Former NASA Astronomer Breaks His Silence About UFOs

by
Marian Rudnyk

A Spacetiki Publications Book

© 2019 Marian Rudnyk.

All text & imagery (including, but not limited to, artwork & photography) contained in this book is copyright ©2017/2019 Marian Rudnyk. All Rights Reserved. No portion of this publication (including, but not limited to all text, photos & artwork may be used (including, but not limited to, transmission, recording, performance, & reproduction) in any way, shape, media or form (whether partial or whole, including digital, internet or any other form), without expressed written permission of author Marian Rudnyk.

Portions of this book are taken from a previously copyrighted unreleased book "INTERSECT (Incident Report On The January 1, 2017 Monrovia, California Multiple-UFO Encounter)" (©2017 Marian Rudnyk) also by Marian Rudnyk, which was never published. This material includes but is not limited to, both text, artwork and photographs, as well as screen capture images that are from the author's original video, which is copyrighted under the name "INTERSECT: The Movie" and includes many of the image elements present in this work. This video is also ©2017 Marian Rudnyk. All Rights Reserved, and is WGA registered as WGA # 1877388.

Spacetiki & all of its related subsidiaries (such as Spacetiki Productions, etc.), its logo, & all related elements, are trademarks of Marian Rudnyk ©1986/2013. "INTERSECT: A Former NASA Astronomer Breaks His Silence About UFOs" & all of its related subsidiaries, logo, & all elements, are trademarks of Marian Rudnyk © 2017/2019.

PAPERBACK PRINT VERSION
First Edition.

ISBN# 978-0-9910474-4-4
WGAW# 2011917
ASIN: B07V661CTS

Designed by Marian Rudnyk
Cover Art & all Interior Book Art & Graphics by Marian Rudnyk TM & © 2019.
Paperback edition contains enhanced content: larger photos & art, & some extra text.

This is a true story. Some names had to be changed, & some redactions made, for privacy & security purposes. Names which provide both the first & last names are authentic. Single name names (with no last name) are all female-name aliases (regardless of the real person's gender) & represent people whose privacy & security is currently protected.

For all the latest news about this & more please make sure to visit:
www.Facebook.com/MarianRudnyk
www.Twitter.com/MarianRudnyk
www.Rudnyk.com

SPACETiKi Books
~Printed in the U.S.A.~

To my brother.

Table Of Contents

Introduction — 1
1: Flashback Perspective..9
   - January 2017: In An Instant Everything Changed... — 9
   - Twenty-Twenty — 11
   - Big Foot Comes To NASA — 11
   - The Gemini Affair — 18
   - The Beginning Of The Full-Circle…Enter Stanton Friedman (Who?) — 24
   - Table Full Of (Lunar) Nuts — 27
   - When Martians Protest You — 39
   - Change Was In The Air — 40
   - Hollywood Here I Come! — 44
   - Close Encounters Of The Hollywood Kind — 49
   - Postscript - From The South Pole To Mars — 54

2: The Main Event - A Firsthand Account..57
   - Some Quick Thoughts — 57
   - And Then It Happened — 57
   - Setting The Stage — 58
   - The Flying Saucers Arrive! — 63
   - What The Pictures Reveal — 77
   - But Wait – There's More…A 5th UFO! — 83
   - What Does It All Mean? — 88
   - In Summary — 99
   - One More For The Road — 99

3: Aftermath: The Mystery Unfolds **103**
   - Timing Is Everything - Almost... — 103
   - Enter U.S. Air Force SPACE COMMAND — 104
   - Decision Time – What Would I Do? — 107
   - But Things Were Now A Whirlwind! — 108
   - Things Escalate & The Military Makes Its Move! — 109
   - The F-18 Fighter Jets Arrived In 2 Groups — 110
   - Reaching Out For Some Answers… — 112

4: And Then The Real "Fun" Began **113**
   - Dreams Versus Fate: Santa Has A Sister, But The UFOs Don't Care — 113
   - Sometimes You Move Back In Order To Move Forward — 114
   - At First It All Started Quietly Enough — 116
   - MIB Goons Crash The Voyager Party — 119
   - And Then Black Friday Turned Black — 126

5: Finally Some Answers..**129**
   - What's In A Name — 129
   - Now It All Makes Sense…! — 130
   - Now Putting It All Together - The BIG Reveal! — 132

6. Some Final Thoughts..**135**

APPENDICES...**151**
APPENDIX 1: Photo Archive — 153
APPENDIX 2: Video Screenshot Gallery — 159
APPENDIX 3: Doing The Math — 171
APPENDIX 4: Support Materials — 179
APPENDIX 5. Original Sighting Stats — 187

Epilogue:...**189**
1. Happy Anniversary Apollo 11! — 189
2. A Special Thank You — 190

Introduction

Intersect? Why "Intersect"? What is an intersect? In math, specifically geometry, it is when two lines cross. Colloquially, it is when one path, or thing, crosses another.

Life is also full of intersects.

When choosing between life paths, we face crossroads. But crossroads are not intersects. When an intersect happens, we are confronted with something that is totally unexpected. It is an event that is so profound, so life changing, that, unlike a crossroad, it provides no choice. It offers no quarter. Whether you accept it or not, it sets you on a new path. The only choice you can make is in how you deal with it.

For me, such an intersect happened on January 1, 2017. I had no idea it was coming. There was no warning. It did not happen by choice. In an instant, when I chose to act on a curiosity, it changed my life, propelling me into a direction I could never have expected nor foretold. It transformed me. If I could go back, perhaps I would undo it – perhaps. Perhaps not. But it's too late for that now.

This is that story. A true story… my story.

One more thing. Before I plunge ahead, I want to be blunt: this is a book about what people commonly call UFOs. If you're cringing and want to put down this book, I don't blame you. But please don't stop just yet. Hear me out…

Not long ago I would have done exactly the same thing – I kid you not. UFO book – I'm outta here. Or, I might even have continued reading (a book such as this), if for no reason, just to be entertained. If I was in the mood, I would devour such a book, just so I could regale my friends with stories of how totally wrong or silly the book was.

I was not alone.

Most scientists, especially astronomers and planetary scientists like me, to this day, feel the same. They suffer from a knee-jerk reaction to automatically reject and summarily dismiss anything having to do with such topics. That included me too.

Working for NASA in multiple capacities, I had been an astronomer, planetary geologist, and imaging specialist. I was trained as an explorer, and challenging the unknown was what I did. Like most other scientists, I considered myself open-minded and curious - eager to solve the mysteries of the Universe. Or so I thought.

Was there life in the Universe? It was a question I often fielded.

I was often requested for speaking engagements. My discoveries as an asteroid hunter, especially made for some surprisingly exciting and interesting tales. For me, the more publicly open the venue, and the more there were all types of people and kids, the better. The events would be lively and invigorating. The questions would be out of the box, challenging – even out of this world! And to me – fun! Kids especially, ask delightfully unpredictable questions that make you think. And that, I always thought, was a wonderful thing.

I recall how in April of 1987 I spoke at the renowned Griffith Observatory in Southern California. I was among the notable scientists who were invited to speak as part of a guest lecture series called "*Solar System Mayhem*". My chosen topic was, "*Mars: The Dead World?*". I spoke at the observatory's historic Griffith Planetarium (now renovated, renamed, and known as the Samuel Oschin Planetarium) – one of the premiere planetariums in the world.

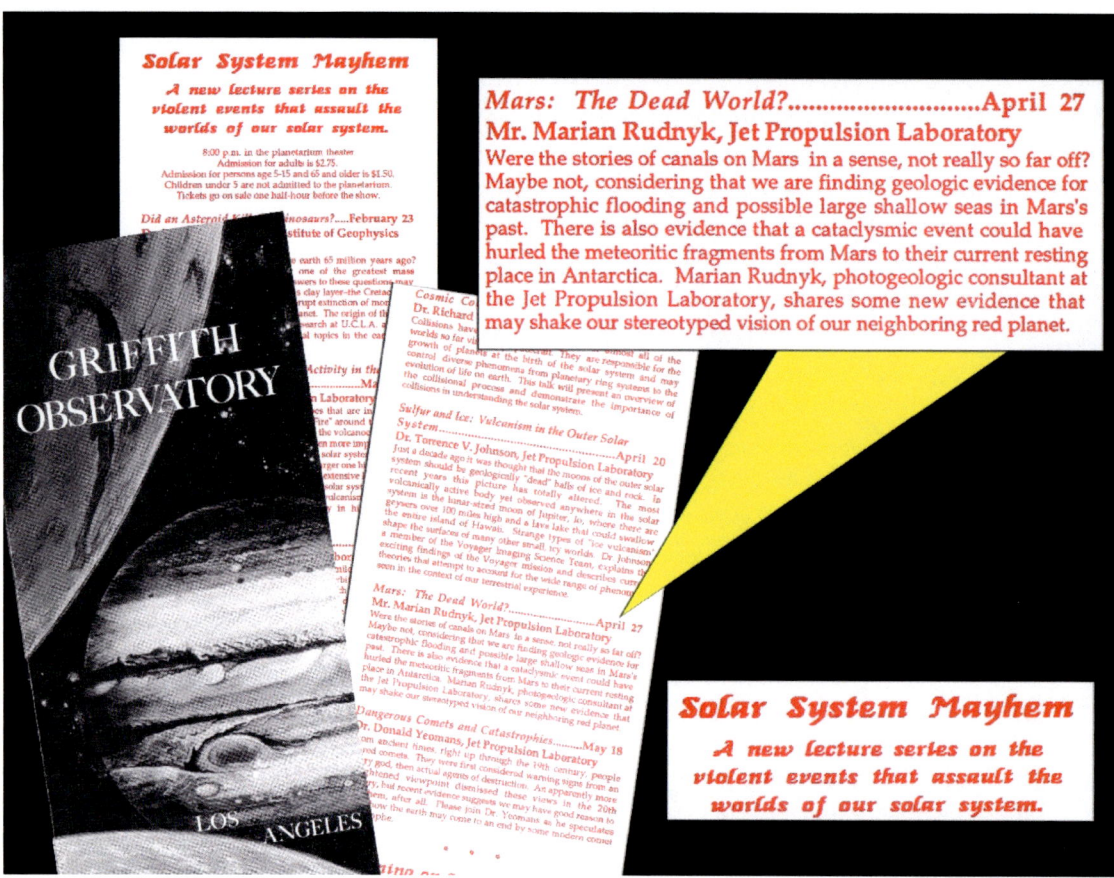

Brochure advertising my 1987 talk "Mars: The Dead World?" – part of Griffith Observatory's sold out hit lecture series on "Solar System Mayhem: A New Lecture Series On The Violent Events That Assault The Worlds Of Our Solar System". (Photo of Personal Archive Collection ©2017 Marian Rudnyk).

Before my presentation I had a quick meet and greet visit with my old friends Observatory Program Supervisor Patrick So and Observatory Director and renowned Archeo-Astronomer, Dr. Edwin Krupp. It was always wonderful meeting with them! Back in the day I used to be a museum guide at the G-O (as we affectionately called Griffith Observatory), and now that I worked as a full-blown astronomer, I could tell that it was not just that we were now professional colleagues, but that they were actually proud of me. And for me, it was always such a treat to return to the place that helped me grow in my career and fondly reminisce about those halcyon days. Afterwards, I quickly went inside.

The planetarium, among the largest in the world, was filled to capacity.

They were a great crowd, and after a lively presentation by me, the questions flew! Especially fun was a large contingent of Boy Scouts who, with scout leaders in tow, surrounded my raised speaking booth and pummeled me with questions after the official Q & A was concluded. They were remiss to ask these questions in an open forum, but mobbing me post-talk, they felt safe. Prime on their minds was what I thought about aliens and UFOs! I was not surprised. This happened more often than you can imagine, and I was always ready, and happy to engage and answer.

I would immediately launch into an animated discussion of the statistical modeling that virtually guaranteed that the Universe was probably teaming with life. I would cite the Drake equation and the virtual statistical inevitability of extraterrestrial life abounding throughout our galaxy. This always made everyone smile. You could see the visions of Star Trek and Star Wars literally fill their eyes! After much discussion of what that life might be like, I always ended with the disclaimer that, although the Universe was indeed probably teaming with life, it was probably not visiting us. Sorry. The distances were just too great. The odds are such a long shot, that any possibility was, well, realistically unreal. And the radio telescopes – I always added in conclusion - they were glaringly silent.

I was warmly comically blunt. There were no little green men. Flying saucers were not landing on the White House lawn and dispensing aliens that demanded, "Take me to your leader." Klatu never threatened our world. Farmers were not being abducted, and cows were not being tipped or ravaged. And don't forget those pesky crop circles. Just as you finished growing a nice crop – BAM! – it was graffiti-covered with cryptic messages from our alien pals - not. You know the spiel. Just soooo much silliness. And you might even be thinking these same things to yourself, right now too. I don't blame you. UFOs? Aliens? Nawww...!

But behind the laughs, you could also see many of these same people kind of cringing. Given a private moment, some brave souls would even venture to privately share with me some sort of weird personal sighting or encounter which, even though they would swear was real, they would still somehow manage to nervously dismiss – as if to excuse themselves of some sort of perceived transgression.

Like I said: I don't blame them, or you, for all the snickers and trepidation.

People like me, scientists no less, filled your minds with every possible reason to close your minds and rationalize away any possibility that anyone might actually be telling you the truth. I didn't see it as a bad thing. I had swallowed the think-speak and was dutifully dispensing the *"Truth"* as it was espoused by all of us at the scientific community. It was canon. It was not to be challenged.

Any talk, of UFOs and aliens, was to be confronted, minds changed, and non-skeptics mocked as not willing to face the truth. If you were scientific you were a skeptic. If you were a "skeptic" you were automatically to be skeptical of UFOs, and never to question accepted "truths". Anything else was to be met as "wild conspiracy" theories wielded by unwitting dupes. And these dupes were to be mocked into cowering submission. There was to be no mercy. Science had now become belief, but you never admitted it.

Additionally, you were to be skeptical of any espoused reasonable explanation that contradicted accepted dogma. You were to rationalize away what anyone supposedly thought they "saw", or thought they "encountered". They were, and still are, told that they "obviously" didn't know what they saw. And then rub it in by saying that even though you weren't there, you still knew better. This so-called *Truth* meant that anything from meteors, balloons, Venus, Jupiter, satellites, the Moon, strange clouds, over-stimulated imaginations, mass hallucinations, fireflies, bugs, swamp gas, and most recently "space dust" – were all responsible for anything that was reported. You never had to prove your explanation, but you made sure that they knew they damned well better prove theirs – already knowing that any facts they brought to the table would be summarily dismissed out of hand – no explanation needed. Remember: you are right. The UFO nut is just that, a nut. And don't let them or anyone within earshot forget it.

That large metallic disk hovering over your house – swamp gas.

A pilot in an airliner and his passengers see a UFO – sorry – it's just space dust.

Those men in black that were following you after you reported seeing something, that's just your paranoia misidentifying over-dressed insurance salesmen. Just ignore the badges, military haircuts, and threats – it all means nothing. It's in your head.

That face on Mars – nothing to see here - just a trick of light and shadow. There's only one picture that looks like a face. Rumors of other pictures existing are just that, rumors.

And the videos of UFOs filmed by those highly trained fighter pilots, with combat experience, and the hundreds of sailors on the ships using radar supporting them – oh, well, they're just following birds, bugs, or balloons. It's "possible" – you are told. Nothing to see here. The pilots are obviously inept – even though we count on them to protect us and secure our country – often on a nuclear level. That they adeptly operate various million-dollar craft that command the land, sea and sky of our planet, suddenly becomes meaningless. They are suddenly reduced to inept uninformed clods that are just arrogant blowhards that make stuff up. Now move along.

No matter how ridiculous the explanation, it is presented as better than the possibility that someone had actually seen something –anything - that was real and out of the ordinary.

Any explanation, no matter how ridiculous, trumped the possibility the UFO was a real unidentified craft.

Any semblance of objectivity went out the window. Am I striking any familiar chords?

I know I am. Like I said, I was no different – only in some ways I was worse: I was on the front lines of being "the problem". (Oddly, I wasn't always this way, but somehow, over time, I had "turned"). I was unapologetic of my condemnation of UFOs and aliens. And although I was always friendly and respectful, I made my position unequivocally clear.

But life is often strange.

Things can change… you encounter an intersect – and suddenly you are ripped away from one side, and thrown to the other side.

Suddenly, what you swore were truths, are now lies – and what you swore were lies, became truth.

This is that story.

This is my story, the story of my transformation.

I didn't look for it. It somehow found me.

This is not fiction - it is fact. Plain and simple, and as you can tell, I'm not going to pull any punches.

Now, for a little perspective let's take a peek at exactly *who* I am. It turns out to be very important – and in a weird twist of fate, now is problematic for me as well now.

I started simply enough in a variety of odd jobs and even managed a movie theatre to help put myself through college. Driven, I often held more than one job and also put in a lot of time doing both heavy, as well as precision, work at industrial high tech factories doing things like machining, manufacturing printed circuit boards, and doing precision assembly of rotary components for both space-based and military components. Happily, near the end of this time period I also had the great pleasure of working as a guide at the famous Griffith Observatory.

All the while I somehow found the time to intern on the side at NASA's Jet Propulsion Laboratory (JPL) for planetary science luminaries Dr. David Pieri (on Europa) and Dr. Tim Parker (on Mars) – and this helped put me at the right place at the right time – and after a few years, and with their kind help, I was in.

At NASA I worked as a planetary geologist, mapping lava flows on Mars and ice fractures on Jupiter's moon Europa. My specialty was Saturn's moon Iapetus. I also managed the NASA-JPL (Jet Propulsion Laboratory) Planetary Image Facility and was on flight teams on such NASA missions as Voyager 2 at Neptune, Magellan at Venus, etc. I specialized in historic space imagery, with special emphasis on NASA's Lunar Orbiters, Mars Viking Landers, the Mariners, both Voyagers and Pioneers, as well as exotic Soviet Venera radar imagery of Venus. While on the Magellan team I worked directly with my Russian counterparts who had been on the Soviet Venera program – made all the more easy because I'm fluent in Ukrainian and also know some Russian. And yes, I know Spanish too.

I was also intimately knowledgeable with the various camera systems on historic missions, and fought tirelessly to preserve not just their data, but also the engineering information and design documentation behind those imaging systems – often confronting and educating ignorant bureaucrats in how such information can provide often-critical context in properly understanding the various data sets. In other words: understanding how a picture was taken helps you understand the pictures themselves - much better. This simple fact was often lost on paper-pushers who only cared about budgets and not the consequences of their cost-cutting shortsighted race to throw out anything "old" – or as they called it "dated". They failed to recognize the importance, in science, of "comparative studies" (comparing old info to new info in order to gain new understandings – how something changes over time can often be a game-changer when it comes to understanding something).

Me working at the massive 48inch Schmidt telescope at Palomar Observatory. Notice the heavy cold weather clothing – back in the day professional astronomers like me worked "at" their telescopes. Because we worked exposed to the elements, and not in warm cushy off-site digital display rooms or offices (often thousands of miles away), we felt personally connected to our work – and that made it all the more personal and special. We had a deep appreciation for our science because we were "hands-on" bonding with it – immersed in it. And we loved it! (Personal Archive Collection ©2017 Marian Rudnyk).

Also while at NASA I did a stint as an astronomer – specifically as an asteroid hunter – and participated in PCAS (Planet Crossing Asteroid Survey), International Halley Watch, INAS (International Near-Earth Asteroid Survey), etc.. I worked at the famous Palomar Observatory and made numerous discoveries and helped develop one of the first automated astronomical film scanning systems (I actually custom-cannibalized an old PDP-1135 industrial scanner that I managed to scrounge up, and converted and re-programmed it to help in the hunt for asteroids – now, of course, modern advanced sleek versions of my experimental Frankensteined-creation are commonplace at nearly every major observatory, and small hi-tech home scanners are ubiquitous).

The first asteroid I discovered, and got to name, was asteroid is 1986 LB, now known as asteroid 4601 Ludkewycz (named in honor of my mom). Most recently I've officially named two more of my discoveries: asteroid 58145 (a.k.a. 1986 PT1) is now asteroid "58145 Noel Kringle" (named in honor of Santa's sister Noel who lives at the South Pole), and asteroid 26090 (a.k.a. 1986 PU1) is now asteroid "26090 Moonlounge" (named after my grade school science hangout). Happily, I still have well as over 100+ others that I still have the fortunate pleasure to name! Unfortunately, the naming process is very deeply flawed, and rooted in ineptitude and insider corruption of the worst kind, which I have been forced to bypass - but that's another story for another time…

The "discovery photo" I took of my first asteroid, 4601 Ludkewycz, which orbits between the orbits of Mars and Jupiter, and is considered a fairly large Main Belt object. It travels around the sun every 4.38 years at a distance of over 224 million miles. I took the photo at Palomar Observatory in 1986 as a timed exposure by tracking on the stars, thus the asteroid (which moved differently from the stars) shows up here as a line. (Personal Archive Collection ©2017 Marian Rudnyk).

I also have done numerous space-related guest lectures at the famous Griffith Observatory in Los Angeles, USC, UCLA, LPSC (Lunar Planetary Science Conference at Johnson Space Center), and countless other universities and schools, scientific conferences, and various events. I've also contributed to numerous scientific publications, written about asteroids for NASA's website, and been published in David Wallechinsky's renowned "People's Almanacs of the 20th Century" books, where I contributed most of the articles about space and planetary science, and space mission history.

During my time with NASA, I was open to the possibility of life in the universe beyond earth, and was even a vocal supporter of efforts by S.E.T.I., but I was never associated with the UFO community, nor UFOlogy as a whole. As a matter of fact, during my time at NASA, I often "tangled" with the fringe elements of the UFO "crowd" – but I'm getting ahead of myself…

So as you can tell by my background, this book is not normal. It is far and beyond any book that is usually about this topic. Astronomers and planetary scientists don't usually openly talk about UFOs. And if they "see" something they certainly don't normally go public with it.

Until now.

I'm tired of the veiled threats, and it's now time for me to speak up.

But I promise you this: I will not be overwhelming you with complex or boring scientific theories, or at the very least I will try my best not to…

So I urge you – please keep reading.

Because this is a very human story.

I will be blunt, truthful and frank in my assessments. Honest in my appraisals. I write this at great risk to my reputation and livelihood. But it shouldn't be that way. The scientific community, even those who share my views privately, will all probably publicly condemn me. Some may even disavow me. Some even warned me they would, saying "Sorry Marian, nothing personal. It's just business." Many in the media, and public at large, may wonder *"what happened"* to me, *"he used to be normal"*. But behind the curtain, you would be very-very surprised at how most of these very same people will actually privately support me, and also tell me that they think this all needs to be "said". Some will even thank me privately, while condemning me publicly. Having heard about the impending release of this book, many already have. Hopefully this book will change all this. Hopefully more scientists will dare to speak the truth. They need to start speaking up too – but that's another story.

Did I mention that I got pictures –AND- a video of my "encounter"? I did – so you're in for a treat! And as of this writing I've already had more sightings. And if you're wondering if the military responded – they did – and still are. The response has shocked even me.

So –please– I urge you to hear me out and read on.

There is a lot more at stake here than just the telling of my story, a lot more at stake for all of us…

1:

Flashback Perspective

January 2017: In An Instant Everything Changed...

...as I walked outside the McDonald's McDiner, I could now clearly see them. My suspicions were correct. As I stood there, I looked up and watched, not one – not two – not even three, but four disk-shaped craft silently move below the low cloud deck.

My eyes widened with the realization: I was watching four flying disks!

They were in a diamond formation – but the last disk was struggling and breaking formation... and descending…

My mind raced: would it crash?

Here I am a NASA astronomer and planetary geologist, standing at a McDonald's McDiner and watching, what are commonly called UFOs, or worse still, colloquially are called "flying saucers" – this was something I could never have imagined my future held for me. Pulling my camera out of my pocket I snapped a series of pictures and a video. Never in my wildest dreams did I realize the consequences of my actions. Scientists don't see flying saucers. And scientists don't take pictures of flying saucers. And scientists most certainly don't also take videos of flying saucers. And to see one struggling – that upped it into the realm of the incredible. It was the perfect storm. There are those that would give anything to witness such an event.

But for me, it was less a blessing, and more of a curse. In time it would haunt me like a never-ending nightmare. There was going to be fallout. There would be consequences. It wasn't going to be pretty. I could have turned around and walked away – but I couldn't. I'm a scientist and observing is one of the things I do. So I stood my ground, observed carefully, and got my footage…

But what I didn't realize as I stood their calmly, with scientific detachment, digitally documenting what I was seeing, was the fact that I was eventually going to be in for a very rude awakening when suddenly I would reluctantly be pulled directly into the twisted churning maelstrom that is the dark world of those who personally witness UFOs – a cruel unforgiving place with strange questions, and even stranger answers. It was going to be like a cold slap in the face that I could only describe as like waking up out of some sort of blurry stupored sleep and into the blinding light of the midday sun and then suddenly *seeing* everything for the first time – in frightening clarity…

So how did I get here? What led me to this point – to this *intersect* of *intersects*?

Was I ever into UFOs? Yes, as a kid. Pictured here are my original childhood copies of Erich Von Däniken's "Chariots Of The Gods" and "Gods From Outer Space", as well as debunker Clifford Wilson's "Crash Go The Chariots". Books I own to this day. By the time I was at NASA all this was barely a distant memory. (Personal Archive Collection Photo ©2019 Marian Rudnyk).

To fully appreciate my predicament, the government attacks, the military muscle-flexing, and everything else I would be forced to endure after my encounter, we need to look back for some perspective. What's even worse was that there would be more encounters to come and it was not going to be pretty. And I've documented all here for the world to see. For me, enough was enough. No more silence. And, yes, a lot of "people" will be angry with me. Too bad. So with all this in mind, I think it's worthwhile, interesting, and important to take a look back first. An awful-awful lot had been happening to bring me to this point. I had been on an incredible journey all along, but I simply didn't realize it until now…

So, lets' start: how did I get here?

How did life propel me to this point where I would be forced to accept what I spent a lifetime considering as unacceptable?

Let's step back in time and put everything into the inescapable context that now has forced me to come to grips with the cold reality that most of the public rarely sees. For me it was an odd journey that culminated in an about-face transformation that I am still, even now, dealing with… struggling with.

Twenty-Twenty

In hindsight, a lot of strange things had happened in the past, especially at NASA, that were to propel me into this new direction, but none stood out more than the collection of events I am about to discuss here in print for the very first time ever. To those that have known me throughout my varied career, many of these events are not only familiar, but represent some of their favorite stories about things I have, let's just say, 'endured'. In many cases, my friends either witnessed these occurrences, and some cases were even a part of them. Because many still work at NASA, or are in DOD related careers, I am forced to protect some of their identities, as needed, to protect them from any possible adverse affects my revelations may cause. The same goes for my Hollywood friends, for my final tale in this chapter is one that reveals my transformation into the role of an actual unwitting 'cover-up conspirator'. At the time, I didn't give it much thought, but in hindsight, it's something that should not have happened. But I'm getting ahead of myself here.

We start this chapter with perhaps one of the seminal events that ever happened to me at NASA. Among the many weird things that have transpired during my time at NASA, none resonates so resoundingly as this first one. It was not actually my first encounter with the UFO fringe, but it was the first big one – and set the stage for the others that followed. In a way it represents a sort of intersect in and of itself, one that propelled me squarely against the UFO community, and set me on a dark path to resist any and all of their efforts at any sort of disclosure. Here we go…

Big Foot Comes To NASA

Enter Eric Beckjord. You've probably never heard of him, few have. Neither did I. Let me preface all this by saying that I had many reservations about including him here in this book, let alone giving him a whole section, but life is like that, sometimes it gives you no choice. And sometimes it can be those choices that confront you and wind up defining you at the intersects in your life. Often such things can lead to disaster. At the time, it nearly was such – but in hindsight – in the long run, it all was actually for the better. But it wasn't until my 2017 UFO encounter that I could put it into proper perspective and realize that.

So just who exactly was Jon-Erik Beckjord? To be blunt, Beckjord was a Big Foot researcher, and self-renowned (in his own deluded mind) expert cryptozoologist who also founded and ran the Cryptozoologic Museum Of Malibu (or C-MOM, as we liked to call it back in the day). The C-MOM, we learned, was nothing more than some rented back rooms of a Malibu beach bar restaurant filled with some ramshackle cryptid displays and photos.

Beckjord's passion was to prove that Big Foots (aka Sasquach, Abominable Snowmen, etc.) were indeed real and actively living in the wild and here among modern society. His theories evolved even to the point that he believed (passionately and with confident, often violent, conviction) that the reason people had a hard time catching any Big Foots was because Big Foots were smarter than we realized and were accessing dimensional doorways that allowed them to easily flee any potential captor. He also spoke of them as being some sort of "trans-dimensional" beings.

Where a man would usually keep a treasured sentimental picture of his wife and/or kids in his wallet, Beckjord had a picture of Big Foot, and would proudly show it off every chance he got. When he first showed it to me, I was too shocked to adequately respond, so I simply gave him a strained smile, and with feigned respect said, "That's nice," as I cringed all the way. It was a warning of things to come, but I never saw it coming.

Now for the record, Beckjord himself could have easily passed for a twisted version of some sort of a northern European Big Foot. He was a massive, towering, and imposing Nordic man with blondish gray hair, who easily stood over six feet tall. He was a physically imposing figure. Beckjord's hair, though straight and thinning, was wild and unkempt, and he had an overall scruffiness to his appearance.

My encounters with him started June 8, 1990 when he burst into my facility, the RPIF (check "*Appendix – 4: Support Materials*" if you want to see Beckjord's signature on my Planetary Image Facility daily sign-in page). To this day I do not know who granted him access. Although the RPIF was one of the very few "public" spaces at JPL (as I had mentioned before), you still needed to get authorization. Perhaps it was somehow through someone who didn't know Beckjord's history, or as a sick cruel joke on me - nevertheless it happened.

Within seconds of our meeting, Beckjord demanded access to everything. I calmed him down and offered to give him a tour so that he could understand where everything was and explained how he could access the facility's vast *then*-world-class collection and resources. The tour was an uneasy affair marred with constant accusations and questions by Beckjord, that attacked my character, as part of some sort of twisted conspiracy by NASA, and the government as a whole, to hide information from him. I assured him that nothing was further from the truth and plowed through the tour. All along he furiously took notes and snapped pictures with his 35mm camera. He made sure nothing escaped his ever-watchful eyes. At the conclusion of the tour – he furiously went to work.

Belligerent, pushy and over-bearing, he promptly created a mess in my facility. For over two weeks he came every day and poured over my various collections and in the process managed to not only thoroughly disrupt the facility's daily operations, but drove away all the various other legitimate scientists and researchers who were trying to do serious work. Among my colleagues and various visiting scientists, Eric Beckjord became known as Eric "Backyard" – or simply "Backyard" - because, as they put it, that's where, as they aptly put it, "he left his brain".

For my part, it was a tough balancing act, but I always tried to be both professional as well as accommodating and respectful. It was a frustrating ordeal. On June 28, 1990, I even gave him a set of spare Viking Mars maps, but these only partially satiated him (check "*Appendix – 4: Support Materials*" if you want to see a copy of the logged Service Request). With every passing day that Beckjord came, he became more and more convinced that he was on the brink of major discoveries, and became ever more demanding and agitated.

In time, though it took a while, I began to somehow understand what was driving him. I surmised that he somehow felt that not only was I hiding and covering up photographic evidence of UFOs and aliens, but he felt strongly that anything he found would later on be "gone". I tried to assure him that this would never be the case. The data and pictures were not going anywhere, and he was welcome to always come back (much to the chagrin of the legitimate scientists also trying to work there – who all were counting the days until he would leave). One of my friends, Jane, an extremely accomplished engineer who was on the Magellan mission to Venus, assisted me in her copious spare time because she took pity on my situation. Unfortunately, for her, she realized too late, to her chagrin, that this was a major mistake. Beckjord said he "found" an image of Guinevere among the cratered terrain on Mars, and he became obsessed with it. Jane valiantly took the brunt of it, as I smiled, thinking I had dodged a bullet, when in reality I had only delayed the inevitable. Beckjord was a formidable force of nature – a raging storm in my facility that scared off other researchers and visiting scientists.

At one point things reached such a fevered crescendo that I tried to calm Beckjord by explaining to him that he could order any of the images he found through the NSSDC (the National Space Science Data Center, which acts to this day as a central purchase and distribution point for various NASA imagery and data, though the advent of the internet has greatly negatively impacted its original vital role). When this proved to not be enough for Beckjord, I finally flippantly said, "Well if you don't trust anyone here, or at NSSDC, then why don't you just take pictures of anything you find here."

To this end he came back the next day with a full photographic setup and dozens of rolls of film, for his 35mm camera - and then spent his remaining days pouring through literally hundreds of photos and maps, and blazing through roll after roll of 35mm film in a furious attempt to photograph everything he could find. Unfortunately, for me, the RPIF often operated on temporary assistant staff, so the brunt of both the administrative and grunt work often all fell to me. That is the seemingly eternal story of the NASA funding situation at JPL, which my friends and associates always stressed was sadly not unique to me, but standard to most of NASA. In the meantime Beckjord's "process" made such a mess that Jane and I struggled to keep up with him. Without Jane's help I would probably still be there, knee-deep in strewn-out data. At least, in hindsight it seemed that way!

Then on the Friday of his last week, what was to be his final day, Beckjord announced to me that he had made groundbreaking discoveries and would be coming back for additional weeks. Although I smiled and congratulated him on his successful "searches", inside I shuddered at the thought of having to baby-sit him through even one more additional day.

With only minutes left in that Friday (we closed at 5pm) Beckjord went nuts and started loudly announcing that he had made a HUGE major discovery! I rushed to him and asked what was wrong (because his excitement often seemed equal parts joy and full on "agitation"). He said that, to the contrary, everything was good, but that he now needed to keep working through the night! I knew this would not be possible for a number of reasons. Prime among them was that our open hours were "the posted hours" – period. Of course, being that it was Friday night, I was also not going to let a fringe lunatic impact a date I had that night with my "rocket-scientist" NASA girlfriend. Additionally, with upper line-management already gone for day, and the Visitor Center entrance closing, there really was no way for him to stay after regular operating hours. NASA-JPL security would certainly not allow it, and I for one supported their adherence to the rules (especially in this case, with my dinner date potentially on the line!).

Yet, in fair deference to him, I still decided to be magnanimous and asked him what it was he found. Maybe, I thought to myself, I could find a way to help him in some other way. I would quickly find out that not only was my kindness misguided, but there was no one that could help Beckjord, not even Beckjord. And soon, I would be the one who would need help – very serious help…

Beckjord, pulled me over to a huge series of photo-binders he had scattered over one of the research tables and ruffled through them, and with a big yelp he exclaimed, "There! See! This is huge!". He pointed to one of the pictures. At his insistence I leaned over and looked at the picture. So what was I looking at? As it turned out, this was one of the binders that contained some of the thousands of Viking Orbiter pictures of Mars. The picture he was pointing to (off hand, I don't recall the actual image number now) but it was of a desert plain with large craters scattered across it – typical Martian terrain.

"What am I looking at?" I asked politely.

He handed me a magnifying glass and pointed to a crater near the center of the image frame. The crater had a series of much smaller craters along one portion of its rim. Interesting – perhaps – but on the whole: unremarkable.

"Ok, I see it," I added calmly, "but what is it about this that has you so interested?" At that, Beckjord verbally exploded!

He accused me of already starting the cover-up and that he wouldn't stand for it. He explained that he now had conclusive proof of life on Mars and that the craters were in the shape of Snow White and the Seven Dwarfs. Was it their faces, or full figures – he wouldn't tell me, saying simply that it "was obvious" and that I was trying to confuse him and it would not work. I looked at the pictures, squinted, and try as I might, I just didn't see it – not anything. True, there was one big crater with seven small ones around it – but that's about as far as any resemblance went. Beckjord would hear none of it! He pounded his fists and demanded access to more pictures of this region.

Here's where the situation escalated, as if the preceding wasn't enough. I calmly explained that I would be happy to look for additional pictures in this region, but it would have to wait until next week. It was Friday, and it was closing time. He would have to come back Monday morning, and I could help him then. At that point Beckjord became even more unhinged and went into a full on ranting rage and threw a tantrum. He yelled at me and made wild accusations. Prime among them was that the minute he would leave, I would be the one who would psychically manipulate the pictures (whatever that means) so that when he returns he would no longer be able to find the pictures. I explained to him that not only did I not have any psychic abilities, I didn't even believe in such things, so he had nothing to worry about.

Beckjord then became even more incensed. He pounded on one of the large heavy wooden tables we use for doing mapping work - it shook with a resounding thud. He demanded he be allowed to stay.

Fearing he might literally snap and physically attack me, I calmly and carefully walked over to my desk, picked up the phone and placed my hand on the key pad. I called JPL Security and asked for assistance. I then walked across the room to Beckjord and told him I'll be happy to see him next week, but that our time today is ended and that Security is on its way to escort him to his car for his own safety since it is after hours.

I walked over to the large metal main door and propped it open.

Beckjord didn't budge. He was livid. His face turned from glowing pink to boiling red. Speaking through his teeth he vowed revenge for denying him his due. Fortunately for me, security must have had someone already in the area, because at that highly charged moment one of the guards walked up and symbolically knocked on the open door and asked if everything was ok. I told him everything was fine (with a hidden worried wink), and asked if he could "please escort Mr. Beckjord to his car" for his own safety. Beckjord was silent. He swept up his various materials into a messy bundle and silently complied. Once he was outside and on his way, with the guard by his side, I called out and wished him goodbye and that I would see him next week. The minute they were out of sight, I quickly shut the door, and rushed to the phone and called security again. I explained in summarized detail what had transpired and that Beckjord posed a possible physical danger and to put the other guards on alert and to please make sure he actually left the JPL premises.

I must have spent an hour just puttering around my facility, cleaning up and trying to calm down. The situation had been so highly charged that I was already dreading Monday. What would happen? Was there anything I could do to somehow literally de-escalate Beckjord? I had the weekend to mull it over. Ultimately, at the wise insistence of my girlfriend at the time, I decided that probably calm heads would prevail and to not over-blow the situation because probably he would calm down over the weekend. Little did I know that nothing would be further from the truth.

Strangely enough, the next week came and went. Nothing happened. No Beckjord. Then the next one. Nothing. And the next after that. Still no Beckjord. After a while things appeared to return to normal and eventually the Beckjord story became an oddball saga that Jane and I would regale to our friends. What we didn't know at the time was that it was just the so-called quiet before the storm.

The first sign that something was wrong was a cryptic, but foreboding, message from my line-management. They will go unnamed here, but suffice it to say that they were a spineless cowardly bunch who worried about their own skins and petty little worlds and had no stomach for standing up to blatant bullying, and often simply withered in the face of any kind of confrontation rather than doing what was right. Their solutions usually revolved around simply firing their way out of anything, regardless of who, or what, was right or correct. I can think of numerous innocents who paid the price for their cowardice. JPL was, Unfortunately, rife with cliques, each with its own competing issues and agendas. Let me set this straight and make this absolutely clear, that the notion that NASA was somehow filled with noble-minded engineers and scientists selflessly doing work for the sake of humanity, is misguided at best – and Unfortunately, in reality, just plain wrong. True, there were a lot of good people there, still are, but often they serve at the whim of some very unsavory characters – paper pushers and bean counters who pretend to understand science and scientists, and sometimes count themselves among their ranks.

In my specific situation, when it came to Beckjord, the squeamish management that was above me was letting me know, without any explanation, that my head was on the chopping block because of my "mistreatment of facility patrons". I could only guess that this was some kind of veiled reference to Beckjord, but they weren't interested in talking to me in the least. They simply were letting me know that they would be rendering a decision and that probably I should be prepared for summary dismissal. The truth, nor my story, did not interest them. In other words: *we're about to fire you – just thought we'd let you know so you would be ready to leave.* Multiple phone calls as well a lot of door-knocking got me nowhere.

Unfortunately, for these people, I had friends in some fairly high places that I could turn to in a pinch. It was time to turn to NASA HQ – and I did. A few phone calls and I quickly found out what had happened. Joe Boyce, the head Program Scientist involved in planetary science groups, including the RPIF, was key in filling me in to what happened, and aided in defending me.

What I learned was shocking. Beckjord had traveled to Washington DC and somehow got a meeting with his senator. It wouldn't be until decades later that I was to find out through friends that Beckjord actually had some military and legal background to fall back on as a resource (and that was how he knew the buttons to push that could cause me maximum "hurt").

In any case, his actions lead to a full-on Senate hearing being convened about NASA *public access abuse* – with me as the fall guy for these supposed abuses. My contacts at NASA HQ were very clear in explaining that I was literally being singled out and that there were those at NASA who were willing to simply fire me so as to make this "problem" go away – with nothing more than a "*Sorry about that Marian, but that's the way it goes. We have programs to protect – you, of course, understand. Greater good, and all that kind of stuff*". The bottom line was that they were scared of Congress. And though they wouldn't admit it, they were just as scared of Beckjord and the trouble he might stir up. Talk of "UFO nuts" was sprinkled among their dialogues.

Fortunately for me, my conversations with Boyce (the head of planetary sciences at NASA HQ at the time) yielded a solution. We called in Dr. Stephen Saunders, who I knew very well. Saunders was a renowned old-time planetary scientist, with a mountain's worth of clout behind him. A deeply respected pillar in the scientific community. Fortunately for me, I had a long and positive history with him that stretched all the way back to the days of my collegiate internships at JPL. And not long ago I had his son do a summer internship at my RPIF (he turned out to be among my best workers – almost as good as my very first assistant Matt – who literally helped me move mountains when we physically relocated the RPIF). So we went way back.

The conversation was a short one. Saunders was a down-to-basics kind of man, and said he had already heard about the incident. He asked me to first explain what had transpired. He wanted my side of things, (though he already knew how this would play out because Beckjord's reputation had preceded him within many in the scientific community). If anything, I guess Beckjord could take pleasure in being known by a lot of scientists – though not in the ways he might have hoped for.

As I explained the painful details to Saunders, I could hear him physically sigh over the phone. I then told him that I felt I had done absolutely nothing wrong, and that I was prepared to fight this – even if it meant defending my actions to a Senate committee. I had extended Beckjord every professional courtesy available to me – and then some. I would not back down. I would fight this in every way possible.

He then asked me if, in light of everything I had just explained, there was a way I could boil down the incident to one simple sentence. I said I sure could: *The RPIF has well-posted operating hours, and we were closing, but JPL security had to be called because Beckjord refused to follow the same rules everyone else followed*. It was that simple. Saunders agreed and said to put the whole affair out of my mind, and assured me he would "make some calls" and it would all just "go away". He told me, "*Marian, you do good work. Important work - with passion and conviction. Forget about this. Just do your job, and now let me do mine. Trust me, this will simply go away. I'll make sure.*" And to his credit, JPL line-management's shock, Beckjord's chagrin, and my eternal gratitude – that is exactly what happened. I not only kept my job, but in the process reinforced my rep as a fighter who would not back down.

There was, however, fallout from all this. Beckjord's heavy-handed onslaught on me had caused me deep resentment against the so-called UFO-nuts. If I had resisted the notion that there was any validity to UFOs before, then this pretty much solidified it. Beckjord's actions were more than just the dismissible antics of a crazed zealot, they nearly cost me my job, and possibly my career. I would not forget. Amazingly, my encounter with Beckjord also took on near legendary proportions among my colleagues and friends. Although it served as fodder for many jokes, it mostly served as a strong cautionary tale about the need to avoid anything or any one who dabbled in the so-called ET/UFO-world, lest you slip and possibly lose your career at their hands. If ever the UFO community needed a lesson of what NOT to do, this was it.

The Gemini Affair

There were a lot of interesting, incredible and absolutely wonderful things that I was a part of at NASA, but there were also many strange and bizarre things as well. Count this one in that latter category.

During my time as consulting planetary scientist and manager at the RPIF I also oversaw its move (as I mentioned before) from the seventh floor of JPL building 264 (at JPL's center) to its southwestern corner in building 202 room 101, right next to the main southern entrance gate. This was not a move I was in favor of, but supposedly "smarter" heads higher up made that decision and it was out of my hands. I decided to make the best of it, and during my tenure there, that location turned out to be fortuitous because I turned lemons into lemonade when I realized I could expand the facility into surrounding rooms as they became available, an leveraged the fact that I was next to an entrance. Inside rumor was that certain higher-ups were trying to "bury" the facility – but because of me, and fortunately for my facility – they failed miserably.

Soon walls came down and I was vastly expanding the place. One of my friends even compared me to the character played by Tony Curtis in the classic Cary Grant war comedy-adventure "Operation Petticoat", where I was scavenging anything and everything around me that wasn't nailed down. And to some extent it was true. Abandoned bookshelves in hallways found their way in, as did gorgeous retro furniture from JPL's NASA heyday of the 50's and 60's, including the beautiful mid-century modern desk I would use that rounded out my office. Massive mural photography, scraped from past missions at places like the NASA warehouse at Cheli, etc. and scheduled for disposal, all made stunning additions to my facility, as did many other "recovered" items. The so-called higher ups had actually wanted to (in what I still think was a very corrupt move) bury the RPIF as a usable/viable facility and move it to a location (to make it less accessible) and thus also shrink it into obscurity, but instead the RPIF went into a massive period of rebirth and began to blossom as never before.

They failed miserably because I managed to bring the RPIF into an era of grandeur it had never known. Confirmation of this came one day when none other than the director of JPL, Dr. Ed Stone, dropped by unexpectedly to chat with me. He had heard that I had an office that was nicer than his, and decided to use our upcoming business as a pretext to check things out personally.

"May I come in?" he said, as I welcomed him.

"Of course," I answered, motioning him inside. "Is everything alright? I was planning on shooting up to your office and seeing you later today."

"Oh, it's fine. But I heard you now have an office that is nicer than mine, so I thought I'd come down your way instead, and take the opportunity to see if all this chatter is correct," he answered as he walked around and checked out my facility.

"So what do you think?" I asked.

"Hmm," is all he said thoughtfully, then finally added, "and your office is where?"

"It's right here," I said pointing, "it's just up this ramp, over there."

He walked up the ramp looked around solemnly, then a big smile slowly crossed his face and he finally said, "Well Marian, I don't know how you managed to pull this off, but it's true, you actually have an office that is nice than mine." Then he added, with a wry smile, "How the hell did you do it? Talk is you're like that guy in that World War 2 movie."

"Operation Petticoat?" I answered, "Yeah, you could say it sort of inspired me."

"As a matter of fact, it is just like that movie," said a wide-eyed Stone, "cuz I recognize that giant Jupiter mural behind your desk. I used to have it back in the day."

"Oh that. Looks good, doesn't it?" I answered with feigned dryness.

"It sure does!" he said, as he shot me an feigned irritated accusatory look. "How the hell did you get it?" he asked in mock outrage.

"It's a long story," I answered with nearly a wink, "I guess that Operation Petticoat story is more true that you may think. I can explain – but you better sit down and take notes." And with that we both laughed and went onto to discuss other subjects at hand.

When we were done and I walked him to the front door, he commented about how cold it was and wondered aloud if that was my way of keeping his visit short, to which I explained that this location had previously been a computer facility, and the extreme cooling was the price I paid for having the nice office, but that I was looking to remedy the situation. "Good luck with that!" he shot back sarcastically, and then added "I can't even manage to order new bookshelves and you think you can change the whole cooling system?"

"Why not?" I countered. "My office is already nicer than yours," and we both laughed again and he went on his way.

As fun, yet trivial, as this anecdote seems at first, it actually proved pivotal. It set the stage, in my mind, to actually try and figure out a way – on my own – to relieve the temperature situation which had escalated to the point to where there were times you could actually see your own breath while there. The Ed Stone encounter brought home to me the fact that I would be largely on my own to solve this. So I set to work.

Now bare with me on this, because it may not seem like it, but we're about to plunge into UFO conspiracy-land.

Now, the main thing about the facility's accommodations were that we had inherited the raised computer floor that the previous occupants had used very effectively to cool all their computers. Just how much space was actually under the floor was a mystery.

One day the mystery of it got the best of me, and one afternoon I plunged into the netherworld below the floor and discovered a whole vast world down there! There was actually some three feet of crawlable height below the floor where you could easily traverse anywhere among all the rooms in a "subterranean" fashion. There were cables and wires and all sorts of stuff that ran here and there. I quickly mapped out the lay of the land of this new "under-floor-world" and proceeded to route some video and other cabling that I needed neatly tucked away, throughout my newfound below-floor expanse. In the coming days my summer hires helped me to completely "re-wire" the facility in amazing fashion.

Then one late afternoon, after everyone in my area had gone home, I decided to explore further into one of the new rooms I had acquired that used to house the remnants of the PDS (Planetary Data System) staff as they had moved out. First, I pulled up part of the carpeting, and then proceeded to pop out one of the floor tiles – which actually were like large heavy panels. With flashlight in hand I plunged into the world below.

I had barely made my way in when a pile of small metallic objects caught my eyes. I made my way over to them and wiped off what looked like decades of dust and dirt. What I saw amazed me – these were vintage metal film canisters! The patina on the metal and labels, as well as their overall appearance made my stomach sink and my hair stand up – these was absolutely NO doubt in my mind: these were authentic – the real deal – the actual original film canisters and films that flew into space on the Gemini missions. These were the originals used by the astronauts during those historic 1960's Gemini missions. The collection was impressive. It included original sets of Gemini Hasselblad flight negatives of hand held pictures taken by the astronauts. There were about a dozen of them, and they represented a hodge-podge of film that made no sense to me. Why this roll, and not another, on this mission, and why some other rolls and not others? It also begged the question, if this represented selected sets, where were the rest? And of course one of the most important and obvious questions: why are original Gemini astronaut-shot photographic rolls hidden under the floor of my facility? Prior to my occupancy, this particular area was home (as I've noted before) to PDS. Nothing PDS did had anything to do with these old manned missions. So many questions, and no answers.

So I set to work. Before it got late I had to authenticate these films beyond the shadow of any doubt. Although my specialty was historic NASA imagery, and I personally had no doubts, I realized that my word alone wouldn't be enough for some people. Which people am I talking about? Those at NASA HQ, and especially those at JSC, where the films were supposed to be archived. These films needed to "go home" – but would anyone believe me? I had to have a way to prove what I had. I knew many of the people at these other NASA centers, and they were often a stubborn lot. Finding stray astronaut pictures so far from their original home would not sit well. They would not only not believe me, but they would consider my story so absurd so as to literally probably not even check.

I quickly poured through vintage technical manuals from the Gemini missions. It was nothing sort of a miracle that I even had the information to prove what I had. Most of these precious resources were supposed to have been intentionally destroyed. I had fought for months the year before to secure their storage at the RPIF. Upper management believed that most were either duplicated elsewhere anyway (without any verification that that was true) or, in their deeply uninformed flawed opinions, was no longer needed. In other words: it's old stuff and no one cares. Ultimately, they told me to toss it, take it, do whatever I wanted with it, but if I left it at the RPIF it had better not impact the space needed for newer unmanned mission imaging data. They especially had no stomach or care for anything from manned missions. I finally was able to protect it by storing it within the confines of my own office area – thus not "officially" impacting the rest of the facility. (Years later I found out that most of these were indeed later destroyed. A rare few were saved because they were forwarded to me by preservationalist-minded people from the Lab. There was little regard for history shown – and this sadly is still the case at JPL. Any current preservation efforts are a pathetic substitute for what truly should happen).

As a quick side note: One particular incident that comes to mind is the Mariner data tapes debacle. These rare analogue tapes were stored in utterly deplorable conditions in a basement. During a particularly heavy time of rains, that whole cluster of poorly underground rooms flooded with water and got into the tapes. When I and several other people authenticated that these not only were the original tapes, but that there was no digitized version that exists – only yellowing old prints of these thousands of pictures of Mars, etc – a quick preservation effort was launched. No one had bothered to even preserve at least a few vintage tape machines to read the pristine data was being threatened with permanent loss. A Lab-wide hunt was launched as well as outreach to other NASA centers, to find old machines as well as people who know how to use them. To the so-called higher ups, the whole issue of preservation represented a double edged sword. On one hand, they couldn't afford to lose the data, especially now that it was revealed to be so severely at risk and with no backups present. Discussions I overheard, sadly, showed a complete lack of care. They would have been content to write the tapes off as a "unfortunate tragic yet unavoidable loss" – but now it was too late. So the flip side of that sword was that they could now be held responsible for any losses. Grudgingly, they cobbled together a preservation effort and converted the tapes to modern digital media before they were lost. Ultimately, the project succeeded and the precious data was saved.

Now I know there might be those of you reading this who might wonder: Why should any of this matter? Why would anyone care about old Mariner pictures of Mars, etc.? The main simple answer is: comparative studies. Where it's Mars, the Moon, or even pictures of Earth, the key to much of the research that people like planetary geologists, climatologists, and others, is to do what compare what was to what is now, and to look for changes. Simply put: changes are the keys to understanding much of what we learn. Without a historic record of "what was", we have no way of comparing to "what is", and therefore, knowing if anything changed, and what such changes might mean.

Now, with all this in mind let's get back to the issue of these Gemini pictures. I checked all the labeling information. A very delicate inspection of the films themselves confirmed what I already knew I knew. It only took minutes for me to authenticate the films as the actual original photographic films the astronauts brought back form space. I was now genuinely super excited! I was holding space exploration history.

Seeing that I still had time, I quickly got on the phone and started making phone calls. Of course, the result was predictable. No one believed me and no one wanted to check actual archives. Worse yet, no one at neither HQ, NSSDC nor JSC would give me a straight answer as to where the originals were supposed to be kept. At the time, I simply felt frustration and anger at what I perceived must be the total lack of organization or responsibility. But the more I thought about it I realize that there were either two choices here: total ineptitude, or some sort of strange cover-up. I discounted the idea of "cover-up" – but that would soon change.

Here is where things take a very dark "conspiracy-esque" turn…

One of the people I had called at JSC and heavily pressured suddenly had called me back. The man was audibly angry – but in a very stern evasive way. His voice betrayed the fact that he was hiding something. He asked me to wait and connected me with someone he wanted me to speak with. This other person was cold, flat and dry. It was like speaking to Sgt. Friday of the old TV show Dragnet. He clearly knew exactly what to ask about. He asked about the labels, and the markings on the film itself. He also asked where I had found the films. He not only seemed "not surprised" but appeared to know what films I specifically had, although I had only provided authentication information to everyone. I never mentioned which specific rolls for each mission I had. He seemed to know.

I tried to ask some of those pressing obvious questions I mentioned before, but he wouldn't answer. Why were these films here at JPL? Why were they hidden under the floor? I found the collection very carefully placed. They were obviously not just "dumped" and clearly could not have simply fallen under the floor – it was impossible. There were no holes in the floor, and the floor panels were heavy and fit together on the frame they sat on very tightly. There was only one scenario: they were intentionally placed there. But why? I asked. All I got was dead silence. As I continued to expound on all this to this man, he cut me off and simply put me on hold. A few tense minutes later he came back and simply said that an overnight special courier would be dispatched. He instructed me to not "talk" and to follow the instructions of the courier. I was to be at my facility at 7 a.m., without question. I was to hand this person everything I found under the floor. He said not to worry about transfer of materials documentation, other "people" would handle the JPL management on my side, and to not discuss anything. He made it very clear to me that I had no choice but to "comply" and that higher-ups he would not name, had already cleared the transfer of the films to his people's custody. Then, without a word, he simply hung up. "How rude!" I thought to myself.

So who was he? What was his name? What was his title? To this day I still don't know. The person I had spoken to originally at JSC had told me to simply follow orders, and it seemed at the time that that was sufficient. I knew that for this special courier to arrive at the door to my facility they would have to get clearance from JPL security, so if at 7am this courier was at my door, then they by default, had to be authentically authorized. As a way to protect myself, just in case, I made some calls to my higher-ups at JPL. Even though they had all already gone home, I left a series of voice messages across the Lab, so as to CYA myself. I would make sure to be at my facility by 6 a.m. so as to give myself time to respond to any inquiries by these people.

That done, there was only one thing to do, call "Jill" (not real name, for protection). I knew Jill very well and Jill was an engineer with deep knowledge of space missions, but also of the "intel-world" and with a side interest on UFOs, though not an outspoken supporter – just someone who was interested in such things and very well informed.

My call to Jill was very short. Jill became hyper excited – and that's saying a lot about Jill, who is normally very "nominal" in their approach to things. As it turned out several of my friends had all been working late that day, and Jill must have spread the word because within minutes they were all arriving at my facility. The excitement was palpable. But this was about more than just historic films. When Jill arrived, Jill explained why the interest and probably why the "mystery man" at JSC was so interested in immediate unquestioning recovery of these films.

What I found out was surprising, and although I wasn't into the who UFO "thing", I had no problem with showing my friends where I found the films and the actual films themselves. Jill explained in great detail that some of the Gemini astronauts had seen UFOs and that some had supposedly taken pictures of them. This was news to me, but I counter-offered. I told my friends, "Well, if there's any UFOs to be found, let's take a look see." Jill then pulled down some scribbled notes and said told me the numbers of the rolls that the UFO community said supposedly should have UFOs on them. As it turned out I had exactly those rolls and a few others Jill hadn't heard of.

Carefully, using archival gloves, we all set about inspecting the films. What we found was astounding, but astounding by not was on the films but what wasn't. As it turned out, Jill may have identified the supposed UFO films rolls, the frames that Jill said contained the UFO images were missing. Each case was the same. Missing frames. At that point a lively debate ensued as to whether "lack of proof indicated proof". As we debated we all proceeded to take turns pouring through all the films, each of us realizing that if nothing else, this was a rare opportunity to handle and view original space-flown films. If nothing else, it was a rare treat for anyone in our various positions, and we knew it.

We all stayed late that night. Nobody wanted to go home. But in the end, I secured the films and we all went home for the night, with all sorts of, what I called nutty UFO conspiracies bouncing around in our heads. I, for one, didn't buy it back then, but at the same time I couldn't shake what Jill called "the obvious".

Jill's case was very potent. Jill bluntly stated the obvious: if there was nothing to hide, then why were these films hidden? And why were the frames that supposedly contained images of UFOs missing? Jill posited that the frames were intentionally removed, and the original films were hidden so as to hide the fact that this had been done. Jill also elaborated that probably what had been done, was that the so-called "master-dupe set" films that the science and NASA community used were shot from these originals in such a way so as to hide the fact that there were any missing frames. Obviously, Jill theorized, certain "intel-people" wanted to preserve the original set, in case they missed anything, so they hid these in a place no one should have found them – but providence had stepped in and placed them squarely in my possession. Jill said I should be very careful and just do whatever the courier says, and that I could bet it had all been cleared at the highest levels. I wasn't really sure I was buying into all this, what I deemed, "conspiracy talk", but I admitted even back then to Jill, I had no other explanation.

The next morning came, and right on time, the courier came. He demanded the films and handed me a form to sign that officially assigned him custody of the films. He was cold and emotionless, but in a very weird threatening way. I complied and he was gone. Later that day, I got several return phone calls from the various higher-ups that I had left messages with. None of them had any answers. The only thing each one said was to ask whether I had complied and handed over the films. When I pressed for answers all were evasive and refused to answer, except to say that this was a non-event and that I shouldn't be concerned.

I found vintage canisters with historic actual space-flown film that contained pictures taken by astronauts, and (as Jill put it), the so-called "good parts" (where supposed UFOs were supposedly visible) were gone. How could I not be curious? And concerned? But as I got busy with the day-to-day grind of my NASA world, the "Gemini Affair" simply turned into a cool party story with no answers. And for the time being, for a number of years, until now, everyone was content to leave it that way. But after my 2017 UFO encounter, it became evident to both Jill and me that there was obviously much more to all this, and I had unwittingly stumbled into the dark shadow world of UFOs and not even realized it.

The Beginning Of The Full-Circle…Enter Stanton Friedman (Who?)
Unlike the nearly unknown Eric Beckjord that I previously discussed, I also got many well-known celebrities from both the scientific as well as the entertainment world. Occasionally, there were also people that were well known to certain communities, but fully unknown to me. This especially included the UFO community. Except for Eric Von Daniken, who I was a fan of as a kid, I knew of nobody else among that group. But that quickly changed one day in 1990.

I got a call from JPL security, at the Main Gate, that they were escorting a guest to me and my facility. He had already been approved by someone among the "big brass" at NASA HQ, and I was being asked to personally attend to his needs. I had barely hung up the phone when there came a knock at the door. The security guard introduced us. I shook his hand (but his name didn't ring a bell), and had him sign in. He was *professorly* in appearance (if that makes any sense), and as we started to talk his already outwardly surly demeanor turned to downright denigrating and demanding. When he asked if I knew who he was, and I apologized but admitted that I did not, he scoffed and muttered something under his breath. I knew better than to ask for clarification, and simply proceeded to try to attend to his needs. He proved to be very knowledgeable with space imagery, though the fact that someone like me, who was obviously much younger, knew just as much – and even more – seemed to irritate him. I think the fact that I was so well versed in imagery, most of which predated me, is what caught him off guard. The fact that I was not just an imaging specialist, but also was adept at the spacecraft engineering systems and was also a planetary geologist, surprised him.

Although he also spent some time on lunar imagery, his main focus was Mars. He was gruff, brusque, dismissive, arrogant and extremely demanding. Although I had other scientists I also had to attend to, he treated me as if he was the only one that mattered. Because it was NASA brass that had gotten him in, I sucked it up, and did my utmost to accommodate his every whim – and there were plenty of them. The fact that at one point his focus was on the Cydonia region of Mars, as well as the Face On Mars, should have been a clue as to who he was. For my part, I was simply way too busy catering to him, and trying to squeeze in everyone else, to care.

As the hours progressed, at one point, he admitted he was impressed by my command of certain data, but then suddenly scoffed, and asked me very pointedly if I had ever discovered anything. When I answered that indeed I had, and explained some of the scientific work I had done, as well as the fact that I had discovered over 200 asteroids, he huffed and went off in muttered scowl, and then proceeded to plunge himself into another volume of Viking Orbiter pictures. I took a deep breath, calmly exhaled, and went back to work to find more of the pictures he needed to see.

When lunchtime approached he surprised me by suddenly thanking me for my help. He explained that he was done, and wanted me to escort him back to the Main Gate. I (gladly) complied.

With him gone, I joined my friends for lunch. As they all talked about the day's events so far, I told them I could top all of that, and proceeded to tell them about my "interesting" guest. When they asked me who he was, I told them that according to my sign-in log he was someone named Stanton Friedman. Suddenly their demeanor changed. They went off on me, demanding, "Don't you know who THAT is?!" "No I don't", answered bluntly, "why, should I? Cuz all I know is that he is a big jerk." They scoffed at me and then proceeded to explain.

Well, to say that Stanton Friedman is both a legend and giant in the UFO community, as well as highly respected among NASA engineers and scientists at large, would be an understatement. I had no idea. Among his many notable accomplishments, he had worked on nuclear propulsion for rockets, no less, which I definitely thought was awesome. However, I shot back at my friends, that as impressive as he sounded, his hyper-focus on Cydonia and the Face On Mars, in my book, put him squarely in the same league as UFO nut-cases such as Richard Hoagland (who my friends had openly mocked with me in the past). They all shot back that he was different and in a class all his own, and wished they had had a chance to meet him.

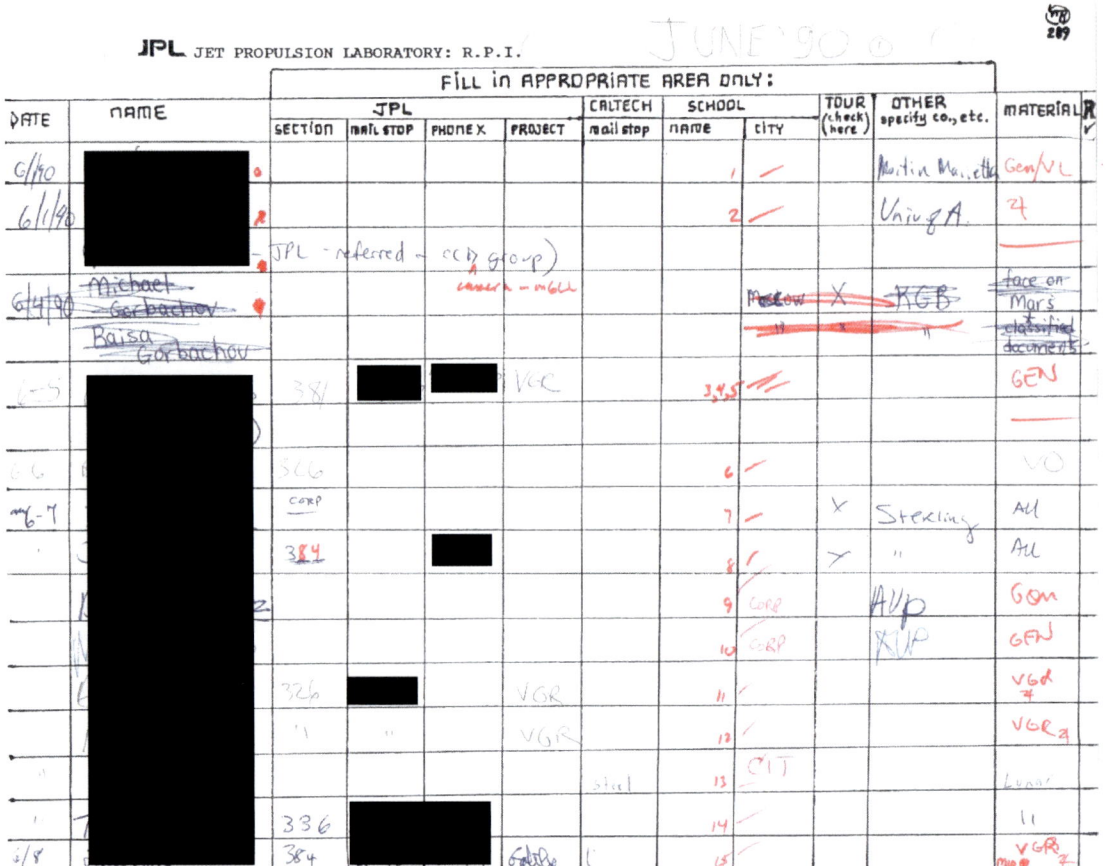

Here is my NASA-JPL RPIF sign in log from June 4, 1990. Notice the crossed out entry, a bit of good-natured Cold War humor with a UFO conspiracy twist. Gorbachev had supposedly come by from Moscow to get copies of Viking Orbiter pictures of the Face On Mars! Funny, yes, but remember, it was an anonymous scientist who wrote in that entry. So as funny as this was, and still is, it also serves as a blunt eye-opening example of what the science community thinks is "silly-tin-foil-hat-garbage" versus what is "real". Note that names and personal info have been redacted for obvious privacy reasons. (Personal Archive Collection Photo ©2019 Marian Rudnyk).

At that point I cut them off and said, "Too bad, cuz he's already gone, so that's the way it goes." Their disappointment, to my surprise, was palpable - but I didn't care. They all hoped he might come back again some time – but me – I was just glad he was gone, and inside just hoped it would stay that way. Little did I know then that things would, in a miraculous and bizarre twist of fate, not only come full circle, but Stanton Friedman would become a friend. For now, let's just say: to be continued later…

Table Full Of (Lunar) Nuts
It was summer 1991 at JPL and after having made good impressions with my contributions to the last couple of JPL Open Houses (also sometimes called "JPL Expos") I was asked to run a large portion of the 1991 JPL Open House, which included not only my RPIF, but all of Section 384, that included prestigious MIPL (the Multi-Mission

Image Processing Lab, later renamed and now known as MIPS, the Multi-Mission Image Processing System). MIPL was the facility where not only some of NASA's best image processing work was done, but also the place that created all those "space visualization" movies that the press and public loved so much. Those included stunning 3D fly-throughs over the surface of Venus, as well as others such dramatically flying over the terrain over the terrain of Miranda (a moon of the planet Uranus) that ends with a smash into the side of a canyon mountain wall. Some of these movies were even presented in 3D using both polarized or anaglyph presentations (depending on the film). These were a huge hit and attracted long lines of people during the Open Houses. Only the RPIF, under my stewardship, had longer lines – which sounds impressive until you realize the crowd control nightmare that suddenly becomes your responsibility. For my part, I got volunteer help from staff from other JPL programs as well as staff and speakers from the renowned Griffith Observatory. Now being put in charge of such a large and important part of the 1991 Open House, I aimed to impress both press and public in ever more ways. My approach was not just about wowing people, but also creating a more seamless coordinated experience that made the whole experience more pleasant and not about spending hours in Disneyland-like mega lines. If I learned anything, it was that when I participated in the 1990 Open House, having between 800-1,200 per hour try to jam into just the RPIF alone was nearly overwhelming – and had I not had the gracious help of the staff from Griffith Observatory (care of Patrick So), I would have been severely overwhelmed.

It was against this backdrop that I decided it was time to modernize. What I wanted was walkie-talkies – and lots of them! I wanted all the key staff under my jurisdiction for the two days of the expo to have walkie-talkies and be in constantly in communication so that we could all act as a real time network and coordinate the flow of people throughout all our exhibits and facilities. As it turned out JPL security actually had a massive cache of walkie-talkies, but it took a lot of wrangling and pushing before the finally agreed to lend us all the units we needed.

Now to people in the twenty-first century, this all seems like a ridiculous task because of the prevalence of cell phones and smartphones in today's day and age. And had cell technology existed the way it does now, I have no doubt that someone would have simply created a "JPL Open House" App – and that would have been the end of that. Problem solved. But this was 1991, and the kind of phone tech that is common now was still nearly a decade away. We needed to accomplish the same thing with walkie-talkies, which to those unfamiliar, are large bulky hand-held radio units still used by some security services, and the like, and they look a lot like today's cellular sat-phones (satellite phones).

As the weekend of the two days of the open house approached, the excitement was building, but we had all the help we had back in 1990, and then some. Additionally, we were all well-rehearsed in how we could use the walkie-talkies to our advantage. But there was one contingency no one foresaw or could prepare for – UFO nuts. And with Mars missions in the works, there was renewed interest in Cydonia and the Face on Mars, and unbeknownst to us, throngs of (what to us were a new "genre") UFO lunar-conspirists.

The new focus among the lunar conspiracy nuts were the Lunar Orbiter pictures from the 1960's. Though we had nothing to hide, for some reason, to these people the existence of the Lunar Orbiter data sets was new and drove them like moths to a flame. To understand what was at stake here, allow me to say a few words about the Lunar Orbiter missions and pictures, because this later becomes key in how I handled the dramatic events that unfolded.

There were actually five Lunar Orbiter missions, and the spacecraft themselves could be roughly described as "declassified" KH-7 spy satellites that were packed with the highest technology available back then. These were "donated" to NASA because newer/better spy satellites were going to be put into service, however, the cameras (from the NRO, and made by Kodak) were downgraded in a way that still met minimum requirements that allowed them to support the then upcoming Apollo moon landing missions, but didn't give away the maximum imaging resolution capability of the original cameras in their original configuration. In other words, the takeaway here is that the military had, and still has, much better equipment than NASA has. If you think NASA was, and still is, leading our tech revolution as it explores space, you'd be wrong. That title belongs to our military, who "share" their highly advanced abilities only when it either suits them, or if they are forced to – and even then, with limitations and caveats. The 1960's Lunar Orbiter program is an excellent example of that.

The camera systems onboard the Lunar Orbiters were complex systems that basically exposed real physical film in order to take pictures. This film was then processed in a special developing system that then scrolled the film past a photomultiplier that scanned it and then transmitted it to Earth via analog signals. Back on Earth, printers printed out this image strips, often colloquially referred to back then as "spaghetti noodles", or just "noodles". These were then literally simply taped together to form giant images panels roughly two by three feet in size. These were then photographed and full-sized prints were then made as well as additional smaller prints of various sizes. Multiple sets were then distributed to numerous NASA facilities, including the RPIFs. The original first generation noodle-montages, the first generation full size giant negatives of the pictures made up of these montages, as well as the first generation prints made of them were all stored at the RPIF at JPL. I know, I often worked with them and led a massive preservation effort to sort and store them all archival standards.

When I took control of the RPIF at JPL I found the Lunar Orbiter materials all in a massive state of disarray. The prints simply were hole-punched and placed in large over-sized binders. These binders and all the other photographic materials had simply been dumped hodgepodge into cardboard boxes and tossed into a storage facility near downtown LA known as "Cheli".

The first time I ever ventured to Cheli I was overwhelmed and saddened by what I saw. It was something like the warehouse at the end of the first "Raiders of the Lost Ark" movie.

The amount of things, especially long-lost historical items was simply staggering. Upon asking, many were slated for destruction. It was a truly sad state of affairs. Often things were simply given away so as to preserve them and save them from destruction from a NASA that simply did not care. And there among these things were the Lunar Orbiter materials. Upon some digging my assistant and I were able to also locate the original cabinets that housed the full sized print collection.

Now you would think that a find like this would be praised, instead my so-called line-management met the discovery with outright disdain. They were more concerned on whether I had the room in my facility to house all this material and what costs may be incurred by housing them. I didn't back down and leveraged my science credentials to run appeals of support from fellow colleagues in the science community who I am to this day grateful to for always taking an interest and stepping up and supporting me. They did so, in this case, in a big way and finally clearances were granted and arrangements for transport to my facility were made. Additionally, I also fit what I could into my own car so as to protect the more fragile materials, such as the negatives.

Of course, when I arrived the morning everything was due to arrive, I found that the delivery had already been made. In typical government fashion, all the boxes had been literally randomly "dumped" all over the sidewalk outside my facility – shockingly unprotected - in the rain!

Long story short, this precious Lunar Orbiter material was not only saved, but I was successful in securing funding for a full archival preservation project. In the end, everything was saved and made available to researchers. (In a sad footnote: the fate of these materials now, as of 2019, is uncertain - the facility was moved several years ago and severely downgraded – and my sources on the inside state that many things were simply tossed out! – but this is still, to me, unconfirmed as of this writing. Regardless, it is a sad state of affairs).

But back then, the Lunar Orbiter data was fully restored and the science community and public gobbled it up. Flash forward a bit now to that JPL open house I was discussing. It was summer 1991, we were armed with walkie-talkies and in just the first day alone the Lab had hosted over 33,000 people. One of the largest attendances ever to-date at that time. As the second (and last) day ramped up we braced for the crowds.

What we couldn't have foreseen was the influx of "UFO nuts" (as we called them). They questioned everything and disrupted some of the displays. The portion of the open house I was in charge of was of particular interest to these people because of all the planetary imagery – this included the lunar material. Although my staff handled everything very professionally and were able to contain most of the situations, one particular one spun wildly out of control. In the end I was being inundated with frantic calls from various staff for me to personally intervene: lunar UFO protesters had climbed onto the display tables in the main quad of JPL and were chanting and demanding "disclosure". I headed out to the location and take charge, but it proved quite the challenge.

The Lunar Orbiter program, which consisted of five extremely successful craft, photographically surveyed some 99% of the moon, which included both near and far sides. The data set produced was massive, especially considering this was with 1960's technology and consisted of 3,062 image frames (each frame being made up of the countless "noodle" strips I have previously discussed). Of these, a whopping 2,180 were high resolution (one meter level). Just assembling them was a Herculean undertaking. The remaining 882 were medium resolution picture frames. In today's hi-def digital world where we can generate terabytes of data at rates that would dwarf the Lunar Orbiter data set, it is important to remember that the surface of the Moon, just like the Earth, is always changing. Where the surface of the Earth is shaped by forces of such as wind, water and geophysically generated events, the predominant processes is predominantly cratering, distantly followed by solar weathering and other exotic processes. Recent scientific data speak of the possibility of lunar quakes, tantalizingly hinting at the possibility that the core of the moon may not really be totally dead and inactive. The data is still pouring in.

Anyway, because the Lunar Orbiter data was created using analog processes, it is much harder to instantly "hide" anything anomalous that the Orbiters may stumble upon. The UFO people, to their credit, were aware of all this and were constantly scouring through these vintage pictures, looking for things that NASA may have missed and supposedly "not yet covered up". And this was the predicament I was now faced with. We were having an open house at JPL, and everyone would be welcome – including the so-called "UFO-nuts" and things were about to get very interesting, very fast.

Now, if you are ever in southern California and have never been to a JPL Expo/Open House, I highly recommend it! The planetary missions may have changed over the years (as a matter of fact they are always changing – so new two Expos are ever really the same). And yes, there are new people there – but as each mission evolves – that is the excitement of planetary science. But JPL has always remained true to one thing - its calling: explore the incredible, often beautiful, and breathtaking universe around us with open unwavering eyes. It's the public's one big chance to get to be face to face with the top planetary scientists in the world, and have often hands-on experiences with the many missions. These open houses usually happen each year at the beginning of summer. But I digress…

That year, was special because we were pulling a lot of firsts and had an aggressive schedule bent on not only impressing people with what we all do at JPL, but immersively blow them away. In the main quad impressive displays welcomed open house guests. From there they could peruse a variety of places, such as the mission control room, and the various planetary and other displays and activities. On our side of things, my team and I had planned an extensive set of activities that took guests through a very wide variety of space-related experiences. A lot of very talented people had really stepped up to make this one of the most memorable JPL Open Houses (often also called Expos) ever, to this day. (For my part I got a Certificate of Appreciation from NASA, see Appendix-4: Support Materials) After those introductory displays in the main quad, we moved people in large groups from venue to venue.

Next up was the spacecraft museum where awed kids and adults alike looked in wide-eyed wonder at the massive interplanetary spacecraft that stood on display. Speakers told the public of the incredible interplanetary journeys that had occurred, were in planning, or were flying even as they spoke. From there they were lined up along a suspended walkway on their way in, in groups, to MIPL (the Multi-mission Image Processing Lab, now actually known as MIPS) where they were awed with 3D planetary sims that our JPL digital wizards had whipped up. From there, they were whisked down to the spacecraft assembly building where flight ready spacecraft were in a massive clean room undergoing final preparations to be shipped to Cape Canaveral for launch to their various planetary destinations. You could watch live from behind glass. From there, they made there way over to the RPIF where various notable speakers, including some amazing people, like my friend Patrick So, had made arrangements to speak from the Griffith Observatory, enthralled people with presentations of exotic planetary imagery both old and new. Then, tours were led through the then world-class exhaustive planetary image collection – then probably the biggest in the world. I had pushed with every fiber of my being to rebuild the facility back to its heyday of the 1970s and far beyond, and now the results were nothing more than spectacular. The public and the press ate it up. We were swamped, as word spread throughout the open house that you gotta make time to check out the planetary image facility (RPIF). We were averaging well over 800 people in the facility per show/tour – well over anything the RPIF had ever done before.

My walkie-talkie crackled with chatter among my staff as they expertly handled the masses, that could have easily overwhelmed such an event in the past, with the skilled ease of orchestra conductors. As I watched the crowds happily move through the RPIF, my mind wandered with making plans for next year. Suddenly, a panicked voice broke my astronomical-tour reverie begging for my: moon base protestors had crashed our JPL Open House! I didn't think such a thing even existed. I clicked my walkie-talkie in speak mode, and asked matter-of-factly, "What's up?"

The reply I got was both shocking and, at the same time, perplexing. "Marian! Help! They're on the tables! They're going nuts!" It was Muriel Lutges, one of my expert helpers from Griffith Observatory. She was working the quad. Well versed in science and astronomy, and expertly adept at handling the public with grace and charm – even she was overwhelmed.
"What do you mean? Someone is "ON" the tables?" I asked, totally perplexed.
"Yes!" declared Muriel. Then another voice chimed in adding, "They're lunar protesters!"
"Marian? Are you there?" cut in one of the scientists at one of our displays in the quad. "You gotta get up here asap! They're on the tables!!"
I asked a friend, who had walked by and overheard the message, and asked her if she knew what all this meant. She said she had no idea.
"Hang tight. I'm on the way right now," I replied with as much calm confidence as I could muster, hoping to help calm everyone - as I bolted out the door.

Not knowing what I was faced with, I ran. Barely a couple of minutes later, I was already at the quad. I stopped and looked around. Before I could take it all in, some of my staff surrounded me, peppering me with comments. "One at a time," I asked. "Just turn around," said one of them. I complied, and was met with a most unsettling sight.

There, on our table was a man and a woman (turned out they were husband and wife), waving some big book and pictures and yelling to the gathering crowds. "NASA is hiding the truth! NASA is lying! There are cities on the moon and we have proof! Make them talk! Stop the deception! Stop the lies! The military are on the moon. There are aliens on the moon too! The aliens are among us, even now! Wake up to the truth…" and on-and-on they ranted.

Crazed screaming lunatics were standing on our tables. Their tirade was loud and relentless. The surrounding crowd was quickly swelling. This had to stop, I thought to myself, before it becomes a disaster. Security had arrived, but I asked them to stand down and let me have a crack at it first. I told them, that it would be better for everybody if I could peacefully diffuse this rather than make a scene with any potentially violent arrests. They agreed, but said they were ready if things went sideways.

I took a deep breath: they want the truth, I thought to myself, so I'll give them the truth. I briskly walked over them, pushing my way through the huge growing throng of listeners, and announced myself to them.

"This man is at the center of the problem!" they instantly declared, and then continued, "He has all the pictures of the planets and the moon. He hides the truth from all of us in his facility. We demand disclosure now…" And on and on they yelled – you get the idea.

Promising them "full disclosure" I managed to calm them and talk them down off the tables. My staff backed me up by immediately dispersing the crowd into the various venues, while I pulled the protesters off to the side to address their concerns.

As it turned out, this couple (protesters from some sort of UFO group focused on lunar cover-ups), were convinced that Lunar Orbiter pictures showed clear evidence of cities on the moon. They showed me the book they were waving around. I asked to see the pictures they had as well. They seemed surprised by my genuine interest and kind attitude and quickly warmed up to me. At that point I walked over to security with them, and assured JPL security that everything was now ok and that I would help these people out.

As security left and everything very rapidly went back to normal, thankfully as if nothing ever happened, I began walking this couple down towards the RPIF. As we walked, I explained the mission of the Planetary Image Facility, and how the doors are always open. I began telling them about our extensive collection, and that everything, including how all the original mission image products were all open and available to not only them, but to everyone in the public. When we arrived at my facility we made our way through the crowds and found a calm corner near the Lunar Orbiter image collection. I told them I would now happily listen, but to not make accusations, and just present to me their facts.

I then had them lay out all their information and listened intently as they detailed their claims.

When they were done, I took a close look at one of the pictures at the center of their claims. It was a Lunar Orbiter image, but I could immediately tell it had been extensively Xeroxed again and again. The image did indeed display curious features: a row of triangular formations that were very obviously artificial. And as grainy and "blown out" as the image was, you could tell the features were really "there" on the film. I told them this and they appeared genuinely pleased and demanded and explanation. At that point I pulled out one of the massive Lunar Orbiter binders. I recognized the image they were focused on. As I flipped the pages I explained that the quality of their pictures was too poor, and told them I would show them a much clearer original. As I found their picture, I laid out the binder, image in full view. As they took it in, I went to another shelf and pulled out one of the thick original 1960s engineering documents. I turned to a section on spacecraft thermal control and waited for them to soak it all in.

"We don't understand?" they stated timidly. "What are we looking at?"

I then explained. The image was a detailed image of the surface of the moon. It was the same one that they had, only much clearer. The picture had the "ribbed" look that is the hallmark of Lunar Orbiter pictures. This is caused by the fact that the Lunar Orbiters used real film when it took its pictures. These were taken in long swaths often referred to, back in their day, as noodles or spaghetti strips. After the film was exposed (picture taken), the film was scrolled over to an mini instant photo lab that developed the film, and then scrolled it past an analog camera that then sent the picture to earth via radio. Back here on earth the signal was "re-interpreted" (processed) as an analog picture and printed out as long negative strips. These strips were then matched, and literally taped together, and the now composite negative (made up of many of these noodles all taped together), was then used to print out a final photograph print of the moon. Of course, once the spacecraft ran out of film, the photographic part of the mission was over.

As arduous and complex as this all sounds, it proved highly effective, and in the mid-1960s five of these Lunar Orbiters went to work and managed to photograph some 99% of the moon in great high resolution detail. Some of the best pictures, and most complete picture sets of the moon that we have to this day. Even as new probes have produced new vast digital image sets, each set, including such vintage sets as Lunar Orbiter, and many others, provide an important snapshot of what the moon looked like at any given time. Remember, the moon's surface is changing and new craters, for example, are constantly appearing as meteors strike it constantly.

As I calmly explained all this to my UFO "friends" I could see they were now riveted and their interest piqued. Amazingly, a small crowd had gathered as I spoke and was now also listening in. I then turned my attention to the so-called elephant in the room – the lunar cities. I then explained that Unfortunately, the explanation for what they thought were cities was rather boring and steeped in engineering, and that although even I thought moon cities would be a cool discovery, these weren't it.

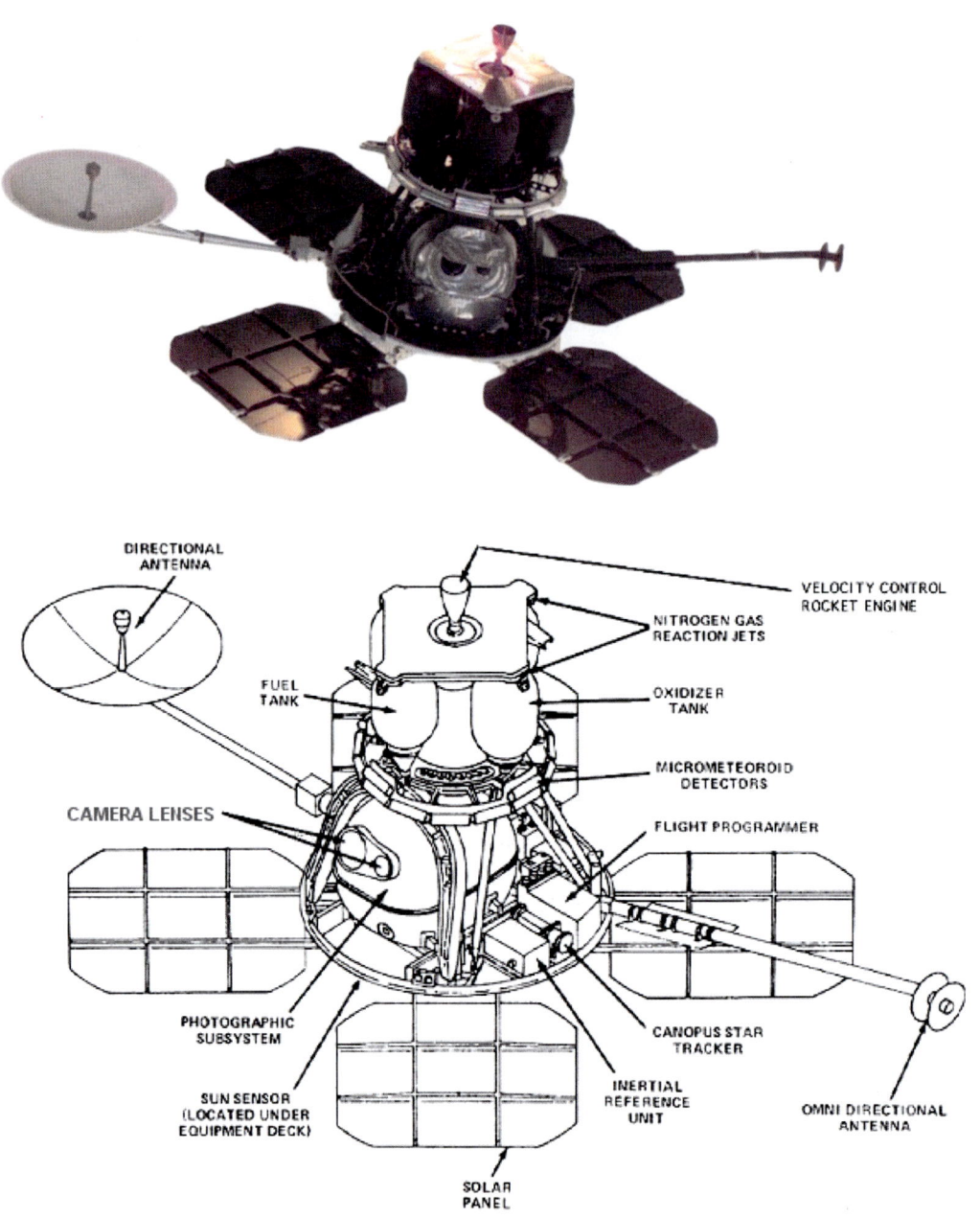

Pictured on the top is Lunar Orbiter 1, and on the bottom is a diagram explaining the components of the spacecraft. On the top picture you can see the cameras peaking out of their spherical housing in the middle of the spacecraft. (Images courtesy of NASA)

I then explained. I asked them to recall how I had just told them about how the film was scrolled through its various processes inside the Lunar Orbiter spacecraft, and elaborated that as it did so such spacecraft often suffered through thermal control problems. As a result, often the sprockets that moved the film sometimes damaged it and caused things commonly called "melt marks". I then began flipping through the dozens of giant images of the moon in just the one binder, and invited them to look.

As everyone leaned in and looked, they could see rows and rows of these marks all over the pictures. I commented, that if these are not melt marks then there are thousands of cities on the moon, so why don't we see them when we look up at the moon? I then pushed the binder over for everyone to flip through for themselves and pulled out a few more so that curious onlookers could also have a look. Lastly, I explained that whoever had distributed the pictures they had, had done them a "disservice" (to put it gently, so as to not insult any of their potential UFO associates) because the pictures had been so heavily copied and recopied via Xerox that it had effectively hidden the true nature of the pictures.

Lastly, I added that the moon was not only an interesting, but a beautiful place, as are all the various planets and objects in our solar system, and that all these pictures were available not only for viewing at the RPIF, but could be purchased directly through NASA's NSSDC (National Space Science Data Center). If anyone needed any assistance with this, I concluded, I would be happy to help. Everyone clapped, and went back to their respective open house activities.

This stunning historic image from Lunar Orbiter 1 is the first ever picture of the Earth taken from lunar orbit. It was taken on August 23, 1966 at 4:36pm GMT. Lunar Orbiter 1 was 236,000 miles (380,000 km.) from Earth at this time, and was 746 miles (1,200 km.) away from/over the Moon's surface. This historic image also vividly displays the "ribbed/noodle" look that is typical of Lunar Orbiter pictures that is indicative of how they were put together with photographic strips called framelets. (NASA Image ID # "LO1-102 (H1, H2, H3)"; NASA SP- 200; NASA Press #L-66-6591; courtesy of NASA)

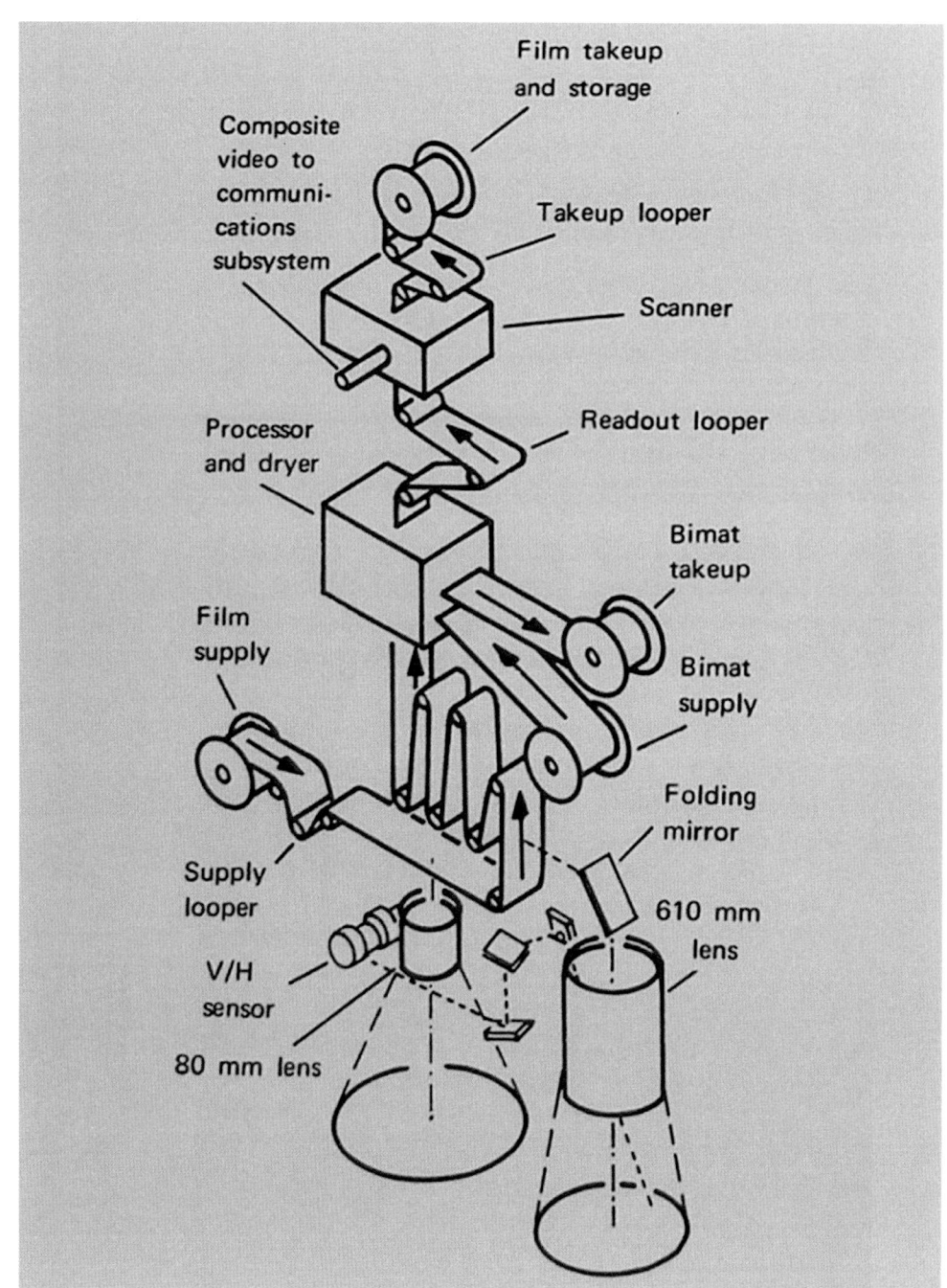

This NASA graphic (on left) details the complex photographic mechanism used by all five of the Lunar Orbiter spacecraft. As can be seen, this was not a digital system, but instead used real film that was exposed, then processed by the spacecraft, and then scanned. Finally, an analog video radio signal was sent and then the images reconstructed back on Earth. As antiquated as this may seem by today's modern digital standards, this analog system proved to not only be extremely robust, but also produced stunning imagery of the moon that is still equal to anything we have today. ("Figure 3: Photographic Susbsystem" from p. 4 from "The Moon As Viewed By Lunar Orbiter"; NASA publication NASA SP-200; 1970; courtesy of NASA).

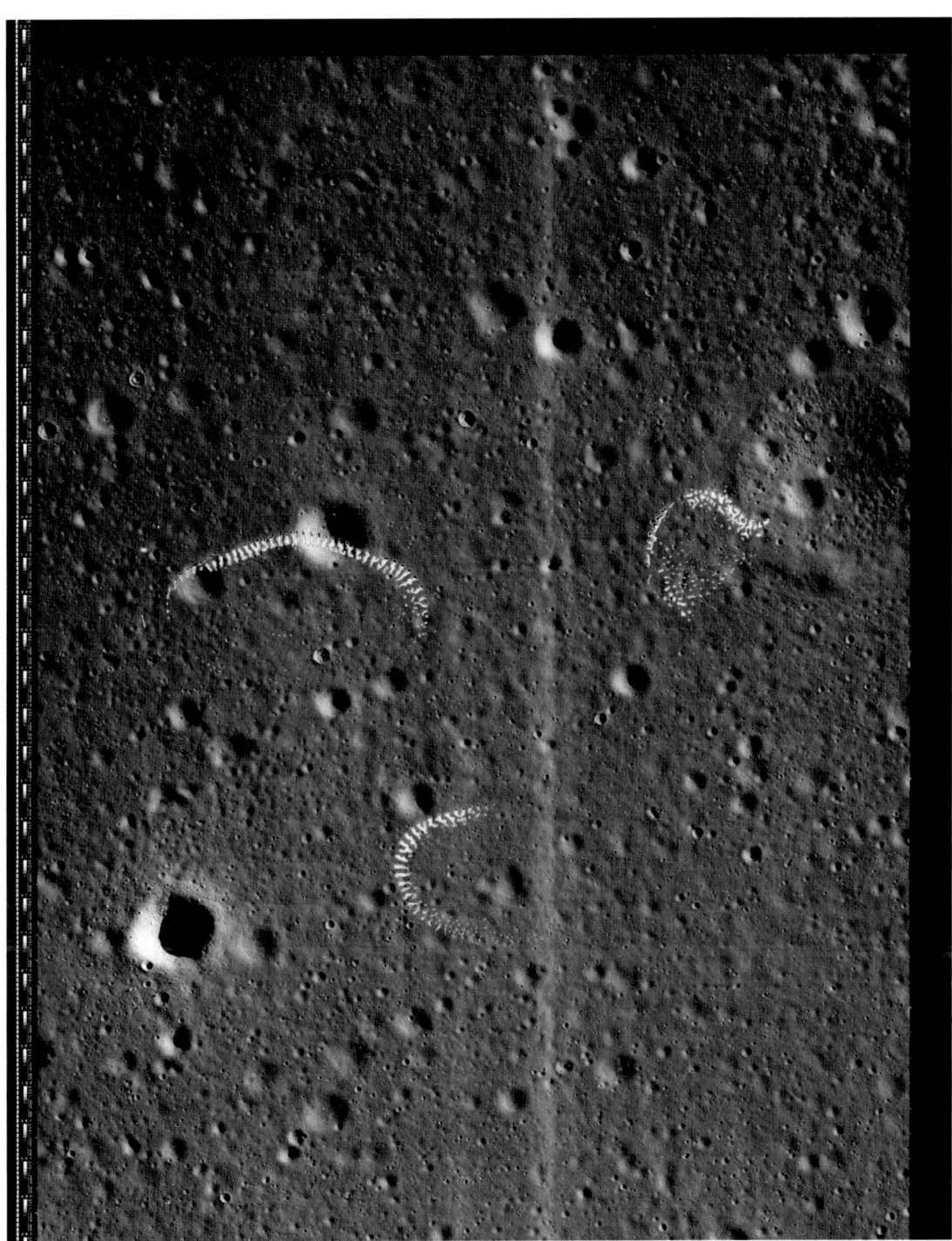

This is image #3166-H3 taken by Lunar Orbiter 3 on February 21, 1967 at 11:59 GMT from an altitude of 33.78 miles (54.37 km). Though I don't recall off hand the exact image that the lunar UFO protesters brought that day of the JPL Expo, I do know this particular image shown here is very much like it. The melt marks that were misidentified as cities by the UFO people are clearly evident here as the mechanically caused thermal/chemical damage splotches they are. The images the protesters had were overly copied close-ups that lacked the context of a full and clear original image such as this one. (Image courtesy of NASA).

As for my "lunar UFO protesters", I could now see that the couple was shaken. Then, incredibly, the woman began to quietly cry. I felt bad for them. Their worldview had been rocked and they needed to reassess what they believed.

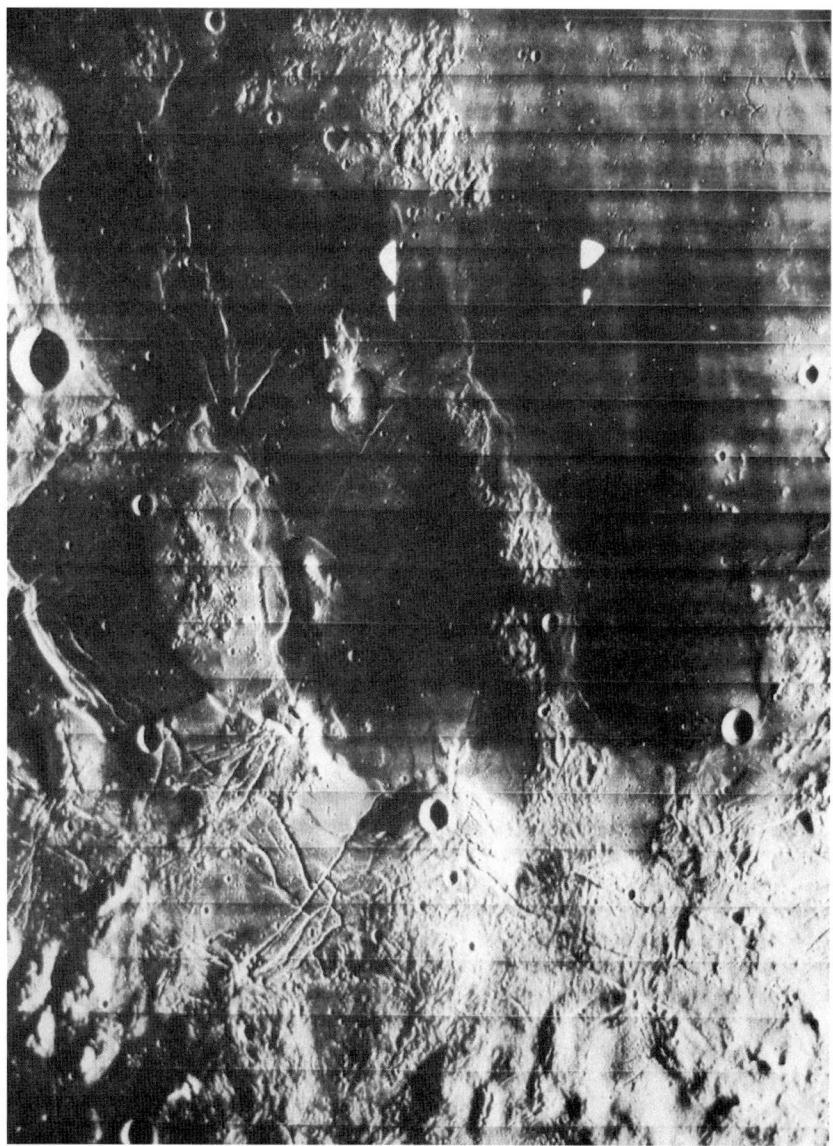

This is Lunar Orbiter 4 image H195 (left) and was taken from an altitude of 2,721 km above the lunar surface. One of the things that the lunar UFO protesters pointed to were triangular cities. Here we see an example of exactly what they were talking about, but in full wide view. We can see that these are processing spots that repeat, and as matter of fact they repeat through a number of images. To objectively drive the point home, I explained that these same areas have been imaged by other spacecraft since then and there are no such spots actually "on" the Moon. Additionally, with a decent telescope you can point it at 24S, 96W on the Moon and see for yourself that these spots are not there (because they are simply film processing blemishes). (NASA SP-200; Image courtesy of NASA).

At that point I briefly interrupted, and comforted them with an invitation. I explained that I had to get back to running the open house, but that here was my business card and that they and their friends were welcome to call me any time and make arrangements to visit whenever they want in the future. The doors of my facility were open to them, just as they are to the whole public. That said, I invited them to enjoy the remainder of the open house and to feel free to reach out to me if they needed anything else.

As they walked away, I could see the husband put his arm gently around his wife and comfort her, as they melted away into the crowd. I have never seen them since.

When Martians Protest You
By the summer of 1992 I had had so many encounters with the so-called UFO crowd that I thought little would surprise me. I had successfully rebuffed them time and again. When it came to stuff like this, many at JPL had come to rely on me and often would send, what I "affectionately" called nut-cases, my way - knowing full well that I would forcefully, yet tactfully, put them in their place. Over the years, Jurrie Van der Woude, the esteemed head of JPL's PAO (Public Affairs Office) had become a very good friend, and even as "tough" as he was, he was fond of simply "shooting these people my way", so that he could deal with "real" public affairs issues rather than this silly nonsense. He knew I, to some extent, relished the challenge of "putting these people in their place". My attitude by this time was, as I previously stated, that little could surprise me. What else could they do? Boy was I wrong!

It was a bright sunny day at JPL and I was working away at the Planetary Image Facility. Suddenly a friend burst through the front doors and excitedly ran up to me – out of breath. As I told him to catch his breath, I asked him what was wrong? What happened? Is he ok?

"It's not me you need to worry about," he answered, "it's you."

"Me?" I answered, perplexed.

"Yeah! Dude, you've got your own protesters! How cool is that!?" he answered with a huge grin.

"What?" I replied. I had heard many outlandish things in my time, but didn't know what to make of this in the least. "What do you mean 'I have my own protesters'? How could I have my own protesters? What the hey?"

My friend then explained that I needed to go outside and check it out. As it turned out the UFO community that I had grown so "fond" of sparing with had upped their game and the ante. I did indeed have my very own protesters. Walking out of my facility in building 202, I walked out past the nearby security gate. This was the south end of JPL. And while 99% of people drove up Oak Grove Drive and straight on to the Main Gate, there was a little noticed turnoff next to the JPL-NASA entrance sign that led to the local fire station (tucked away among the pine trees in that area) and just beyond that, a South Gate.

By walking up that road I could approach that main sign from the side and behind, and for the most part, be largely unnoticed. As I did so I didn't have to go all the way up to get my answer. There, in front of the main JPL-NASA sign was a giant mob of highly organized protestors. They were chanting and yelling at passing vehicles that were coming in and out of JPL and waving signs and banners – many NAMING ME! They attacked me as being at the forefront of a massive cover-up. The Face On Mars was real – I was deemed a Truth-denier. Mars data that proved there were aliens? Covered-up – my fault! Pictures of UFOs that were taken by astronauts – hidden by me! The attacks were loud, ugly and relentless. They demanded answers. They demanded my head. I had seen enough. I quickly, and carefully, made my way back to the safety of my facility.

Back at my desk my phone rang off the hook. The views of my friends was more amusement than anything else, and the attitude of *"better you than me – ha-ha-ha!"* The higher-ups were not amused as word spread up the chain all the way to NASA-HQ – where, fortunately for me, calmer and cooler heads prevailed. The head of the planetary program there, Joy Boyce, told me not to worry. He would make sure to deflect things on his end, and for me to just lay low and this would all blow over. He explained that these things occasionally happen, and not to worry, and take it as a "badge of honor" because they actually consider you knowledgeable enough to protest against.

By week's end the protesters had finally left, and as things eventually settled down, my reputation as someone strong enough to stand up against outside forces was beyond cemented. I knew even back then, that there was a lot that could have gone wrong, but that fortunately for me I not only handled myself well, I was also very lucky to have many people who had my back. Among my many colleagues the story took on a sort of badge-of-honor legendary proportion – kind of like something that was scary at the time, but now you can laugh about it. For me personally, it was a very surreal event, and it made it very clear to me: the UFO community were now the "enemy" – and I would now make sure to, with great care, treat them as such.

Change Was In The Air
After everything that had happened at NASA, I had had enough.

When the Mars Observer spacecraft was tragically lost in route to Mars, nearly 200 jobs went with it. JPL was literally rocked and people were understandably upset and scrambling to find work and survive. Fortunately, I was not part of the pink slip crowd, but my budget was still severely impacted.

Additionally, the advent of the internet was beginning to take its toll. Although we didn't know it yet, soon the world would be changed by the advent of Windows 95 and new revolutions in personal computing and communications. With the early Mac machines popping up all around the Lab, and JEMS (JPL Electronic Mail System) and various digital media permeating everything – change was in the air.

Within this climate of drastic digital change and budget cuts, came the rise of a new line of thinking at NASA: the "faster, better, cheaper" mantra. Unfortunately, this approach also spawned "evil" spin-offs. One of these were things like the view that libraries, bookstores, and even things like books themselves, should now be considered things of the past. Everything should be put on CD's. Soon even the digital disks themselves would be scrapped in favor of going fully online-only. Parallel to all this came a new line of thinking at not only JPL, but throughout the NASA-verse: data facilities of all types are relics of the past. As we now know, things like bookstores, libraries and even data facilities have actually made a resurgence and found their own effective niche. But back then, everyone was hit hard. Everyone wanted to get on the digital bandwagon, and anything that "stank" of analog was to be replaced or abandoned, and anyone who stood in the way, pushed aside as prehistoric. For me, it meant I was fighting a seemingly never-ending battle to preserve historic materials and space photography. The "science" was being sucked out of my facility, replaced by a constant panicked scrambling to just survive.

Interestingly, one of the things that had followed me throughout my many varied career at NASA was my writing and drawing talents. I finally copyrighted and trademarked my much beloved comic strip, called "Lunar Loonies", that I created as a kid. It had been a longtime favorite at the Lab (JPL-NASA). It featured a pair of alien crater creatures named Dot and Dixie that lived on the moon and traveled freely all over outer space, visiting all sorts of planets and moons, etc. along the way. They had had a very successful run in my university newspaper, and in many local publications. At JPL there were even people who sported Dot-&-Dixie tee shirts! These had been made for me, and everybody else, by fans of the characters. The main receptionist for all of JPL, at the main entrance, Bobbie Mann, a jovial woman who was probably one of the friendliest people you could ever meet on the planet, was one of my biggest fans. So much so that when she had her wedding (at Caltech, no less, and with all the biggest NASA brass also in attendance), I was also invited - & I made a special Dot-&-Dixie wedding gift just for her.

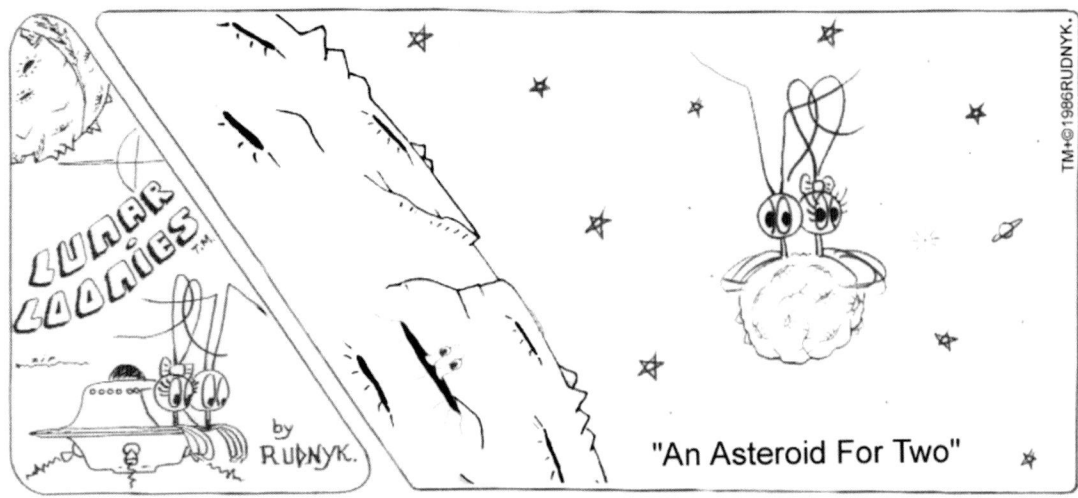

My much beloved comic strip, "Lunar Loonies", often known colloquially simply as "Dot & Dixie" because the two main characters (seen here) are, respectively, Dot and his girlfriend Dixie. They are friendly aliens native to our moon - known collectively as 'crater creatures'. The strip features a whole host of various characters including other aliens, various visiting astronauts and even talking craters & meteors. The strip continues to circulate privately and I am in talks to possibly finally get it syndicated, as well as exploring other options.(TM & ©1986/2019 Marian Rudnyk)

Those were halcyon days at JPL, and NASA as a whole. But Unfortunately, things were changing – rapidly – and not for the better. Widespread budget cuts and massive layoffs were looming on the horizon. Already many departments had been downsized. My friends and colleagues were all over me to explore my creative side. "We're stuck here, but you're not," they would say. "You have some really cool talents and it's time you pursue them. This is your opportunity." I knew I had accomplished most of the goals I had set for myself there, but still, it the thought of leaving was hard to stomach. But as events rapidly unfolded I could see that NASA was no longer the same NASA I had grown up with and loved so much. It was then that I realized that, hard as it was, it was indeed time for a change.

What was especially disheartening during those times was that NASA seemed to have lost its way. Apollo was cancelled. The space shuttle, while a really useful, and I feel, a necessary tool, had become an end unto itself. And the US space station kept getting renamed (Alpha, Freedom, etc.) as it was simultaneously continuously being scaled down. Even after it was severely stripped down from its grand original design, NASA still blinked and farmed out sections of it to a new consortium of international partners – which, in a sad ironic twist, included our enemies – the Russians. This final version, the International Space Station, or ISS, cemented NASA's path for the last over 40 years, often colloquially widely dubbed "*the road to nowhere*". At the end of the 1980's and throughout 90's the US manned space program languished in LEO (low-Earth orbit). And as magnificent a flying machine as the space shuttle was, with the exception of the rare Hubble telescope repair mission or the like, its full capability was mostly wasted until it was Ultimately, cancelled in the 2000's. In doing so NASA then literally surrendered its lead as it opted to rely on primitive Russian Soyuz rockets to fly them to the ISS. Although the government and NASA tried to spin all this as moving forward, the reality was an overwhelming feeling of humiliation and of being trapped on a rudderless ship.

During this time of what I like to call the "Great Slide" for NASA, the rift between the manned and unmanned space programs began to widen to the extreme. Previously the unmanned robotic missions pushed the boundaries of our exploration as a way to both explore and to pave the way for manned missions to follow in their wake. But by the late 1980's and 1990's the robotic missions pushed on with great success and as an end to themselves. Missions like the Pioneer 10 & 11, Voyager 1 & 2, Galileo, Cassini and other missions to the outer planets, as well as a slew of other missions including go to Pluto, comets, asteroids, and a return to Mercury & Mars – the unmanned space program seemed on fire! But nothing could be further from the truth.

Yes, these missions were incredible and necessary and wonderful – but NASA still lacked balance and direction. Most of these missions had their roots in approvals and designs that often went back decades. The fights over new missions were often bitter and ugly, and often scientists resorted to working on portions of them in their home living rooms (as was the case with Mars Pathfinder). Funding battles seemed to undercut everything. Mantras such as "smaller, faster, better, cheaper", came, failed and went. While the science from these missions was spectacular, most of the public was jaded and largely unaware of their incredible scope and magnitude. So instead of inspiring further manned missions to build on this, it was many in the robotic part of the space agency that pushed against any sort of manned program. Many argued we should only send robotic craft out, and that sending people out to explore space was now some sort of passé dated notion.

I remember in the late spring of 1992 how management dispatched people around the Lab to collect signatures for a petition. Ultimately, I was also approached. The petition was a plea by JPL for NASA to severely scale back, or even eliminate, funding for manned space programs, in favor of saving various JPL unmanned missions from the funding "chopping block". Their focus was the space station. They wanted it killed off. When I refused to sign and urged my staff to do the same, I received all level of threats, including losing my job for my supposed "disloyalty". In reality, all I advocated (and still do) was a balanced approach that harkened back to NASA's original way of doing business: with manned and unmanned programs working together in unison in a focused fashion on one vision – one unified mission. Ultimately, I never signed and the threats went away, but it was now abundantly clear to me that the so-called writing was indeed on the wall.

So by the end of 1992, I had a decision to make: do I continue on in the sciences at JPL, or perhaps evolve into something else at NASA. I had options. Although I didn't want to go back to asteroid hunting with Glo Helin (for a variety of reasons better left to another book), I had offers to join the Galileo Jupiter mission team, as well as the fledgling Cassini mission to Saturn team. Part of me was torn, but part of me knew the answer.

Additionally, I had already made it to the last group of 100 once for the space shuttle astronaut program in the past, so also I pondered trying again. I had succeeded at becoming an astronomer and planetary geologist, but being an astronaut was still an unfulfilled dream. Unfortunately, I had serious doubts about "where" NASA was headed. Although I was a big supporter of the space shuttle program, and considered the shuttle a beautiful craft and an important and necessary technological marvel, like many others, I felt that by just staying in Earth orbit, NASA had lost its vision and passion for real exploration. Going back to the Moon and onwards to Mars – soon – I felt then (and still do) – is vital and critical. The constant infighting between manned and unmanned programs at NASA didn't help NASA – but it help me make a decision by pushing me away.

Additionally, as I had previously pointed out, I felt it was important that unmanned missions should serve to Ultimately, support manned efforts, and not simply become an end to themselves. If at the present time, NASA wasn't doing any real exploration; well then, I had my answer. It was time to move on and address my other "side" – my creative side. The writing was on the wall, having worked extensively in the sciences, it was time for me to go and address my artistic talents.

Like the many friends before me, I decide to join the ranks of those that somehow leveraged their space skills in… Hollywood!

Hollywood Here I Come!
A career change of this magnitude was massive. I spent years at nearly a dozen part-time jobs as I carved out my own retraining path. I found a way to take a huge variety of art and animation classes offered through the Animation Union (IATSE Local 839). Every semester the Union had a few open seats available to artists outside of the union – I scrambled for every one I could get, and vowed to myself, that if this worked and I could actually land an animation job, then I would proudly join this union for helping me reach this new dream.

After years of art classes, I was ready and my girlfriend at the time (an independent film producer) finally felt I was ready and helped my assemble my artist portfolio and shop it around Hollywood. These were very hard, and yet also fun and carefree times. Anything and everything seemed possible. All the UFO events of my NASA past seemed far behind me, but little did I know that nothing was further from the truth.

It was at this critical juncture, another intersect, that two breaks happened back-to-back in rapid fashion. It was 1996 and I got a frantic call from my girlfriend: she had spoken to a contact at ACME Filmworks (who happened to be next door to the production gig she was on) and based on her and some mutual friends' recommendations, ACME was willing to bring me onboard as a traditional animator on a project for Levis Jeans For Women commercial called "Trading Places" – sight unseen! I would be working directly with world famous animator Raimund Krumme. The catch? I had to be there, at their Hollywood studios, within the hour and be ready to immediately start working.

Throwing myself together, I tossed my portfolio into the trunk of my 1968 Buick Skylark and headed out. In a whirlwind that would make even the most experienced artist's head spin, I instantly found myself at a traditional animation desk being fed key frames for the commercial directly from Raimund Krumme himself. I immediately set out cranking out in-betweens and additional key frames based on the various materials that were being provided me. An hour later a group of people invaded my space – took the stack of artwork I had cranked out and told me to sit tight. They were going to scan the artwork, run an animatic test, and let me know how I did. I didn't exactly understand but I waited. After a bit they ushered me into a room with a computer monitor, all smiles and said take a look. There was my artwork – alive in motion on the screen. In typical newbie fashion, I asked if they needed to see my portfolio, at which point they all exploded with laughter telling me, "You're kidding, right? Take a look at your work. It's fantastic – you're in!" And with that they sent me back to work. The job was mine. That commercial aired later that year and even won awards.

Photograph (left) of a promotional postcard that promoted Levi Strauss & Company's Levi's Jeans For Women television ad campaign. It features a scene from the animated TV commercial (© Levi Strauss & Company) on which I worked. (Photo of postcard from personal archive collection ©2019 Marian Rudnyk)

As 1997 began, I was still riding high from stint at ACME, when I got a call from my dad. He was very excited and asked me to come right over because he had something to show me. It was a small article in a back page of the newspaper. It was calling for "displaced aerospace workers" to apply for a new California state funded re-training program to find people like me jobs in Hollywood. It turned out to be one of the most incredible breaks I had had to date. I slapped together a quick resume and faxed it over to the people running the program that same day. Shockingly, later that night I already got a phone call! The person on the phone was visibly excited (even over the phone) & yet oddly simultaneously stern. He explained that I was a perfect match for the program and was blown away by my science background, and then proceeded to "lay down the law", so to speak: I was to come in to their offices that Friday. I would see dozens of people in the waiting room, but I was NOT to talk with them and not tell anyone why I was there. The reason? They were all still going through an extensive screening program. Only 8 people were going to make it out of all the countless people applying in the whole state. By I was different, they told me. I was simply "in" – period. I got the first and top spot – but with the condition I didn't say anything to anyone else. They wanted me to simply come in for a quick meet-&-greet and sign some papers and explain the program to me.

I didn't know what to expect, but obviously I was very excited. I was told that when I arrived to avoid everyone else, and simply announce myself to the lady at the front desk and she would handle me. As I entered the main waiting room I was overwhelmed, the room was beyond packed. As I looked around trying to get my bearings I bumped into a guy named Rimas Juchnevicius. How do I remember such a complicated name, because he gave me his business card – a card like no other. It was very unique in that on the flip side of his card was a pronunciation guide! Yes, Rimas was quite a character, and unbeknownst to me back then, we would be working together later on a number of Hollywood projects in the future. But that was still to come. For now he was excited and detailed the stages of interviews he had endured and how he hoped he would qualify. I evaded any mention of my own situation. At a convenient moment, I excused myself and made my way through all the people and found the front desk – actually a table with a very busy woman coordinating everything. As I approached, she barely gave me a notice – at least until I announced my name – at which point she excitedly introduced herself and then with a smile, leaned over and whispered, "You're expected. Follow me".

I was ushered into a room where a number of people sitting behind several tables pushed together, sitting there as if they were at a discussion panel. She introduced me, and then each of the people took their turn introducing themselves. Many were supposedly top people in Hollywood, but this was not yet my world, and although some names, or faces, seemed familiar, I didn't really know them, so I politely smiled knowingly as each one introduced themselves and shook my hand. They then took turns explaining the program to me. I would report to a studio were we would be trained for one month. We would be working on top of the line Hollywood computer systems learning various visual effect techniques, including digital painting. At the conclusion of the month we would walk out with not just our training, but a VHS tape demo reel of the best work we would produce while there so as to showcase our newfound abilities. At that point we would be on our own to get a job at a major Hollywood, and they would try to help generate leads for us to give us a leg up.

This all sounded very exciting and I thanked them and told them I was very eager to start. At that point, a thin woman who everyone seemed to treat as being some very "important" Hollywood type who was involved in this pushed a piece of paper across the table towards me and simply said, "You need to sign this." I took the pen and signed. Then she leaned way over the large table towards me, and looked me deep in the eyes with an almost menacing look, and I will never forget her words as she said very directly, "That's it. You're simply in. No one else is. You are not to say anything about this to anyone else as you leave." I nodded that I understood. Then she continued, "We're very glad to have you onboard. You're our number one pick." Then with an almost cold command she finished with, "So don't disappoint us".

"I won't. Don't worry. Thank you. I won't," is all I said as I smiled, shook her hand, and left the room.

That month of training went by in a flash. There were only 8 of us. Rimas had made it and was in the group. We were told that if we did well then 8 more people would be chosen for a second group, and then this pilot program would end and be evaluated. It was stressed that we needed to do well for this process to proceed successfully.

The training was fascinating. I literally devoured everything they showed me. I quickly discovered that having had several years of union art classes and that animation job on that commercial gave me a distinct edge – and that was a good thing – not only for me, but for the other 7 people since I tried to share what I could of what I had already experienced.

Fast forward to a week after completing this one-month program and I got a call to interview for James Cameron's new film… "Titanic"! Nailing the interview, a week after that I was actually working as a full fledged Hollywood visual effects artist – on "Titanic". As phenomenal as that seems nowadays in hindsight, back then "Titanic" was a way-way-over-budget production that was expected to flop so badly that all the long term career artists that were already there were telling us "newbies" to not get our hopes up, and just be grateful we were on a big Hollywood production and to simply at least try to learn as much as we could from it. Fortunately for me I worked so hard that I shined and was one of the last ones still working on some of the last shots of the film. As November rolled around, I found myself still cranking out extra last minute shots during the Thanksgiving holiday. Even though it was my first ever Hollywood film I did so well that I was told that my name would be in the credits, something not common when one is just starting out.

Pictured here is the LA Times newspaper, San Gabriel Valley edition, from March 20, 1998, with a cover article about me – one of many back then. At the time, my career move from NASA astronomer to Hollywood animator and visual effects artist – basically from scientist to artist – sparked peoples' imaginations and interest, heralding it as "A 'Titanic' Career Move". Working on Titanic proved to be a lot of hard, but fun, work coupled with a non-stop whirlwind schedule of press and media interviews. Truly it was one dream come true after another… and then another and another. (Photo of newspaper from personal archive collection ©2019 Marian Rudnyk.)

Of course, "Titanic" was released very shortly after I wrapped in December of 1997, and made history and helped launch my Hollywood career in a massive ways I could never have predicted in my wildest dreams! Every door in Hollywood was opened to me and dove in, bouncing from one big Hollywood studio to another and in the process working on one big movie after another. I was flying high and loving it. Among my many early projects was a stint as Science Consultant on Armageddon, where my assignment was specifically to address the look and feel of all the meteors that bombarded the earth, especially in the early part of the film. One of the things I successfully convinced the 3D CGI team to do is not having boring vertical asteroid strikes (which was what the director Michael Bay originally called for). Showing them scientific literature that backed me, I was able to show that about 20% of meteors come in a shallow (near horizontal angles) of less than 25 degrees. This, I explained, could make for much more exciting scenes where meteors could literally be shown punching through multiple buildings before finally hitting the ground. Another piece of scientific reality I was able to inject was the idea of concussion waves. What this meant was that as a meteor hurtled across the sky, there was a concussion wave in front of it, of air that was being compressed ahead of each incoming meteor. At first the CGI team balked at the relevance of this tidbit of physics, but when I explained the artistic fun they could have with this, they were elated. One good example of this effect was the scene near the start of the movie where a meteor strikes New York's Grand Central Station. Watch closely and you'll notice that in the first split second all the windows shatter and blow in dramatically just before the meteor itself tears through the station – thus adding an extra element of excitement to the scene. Thus, in a film often called exciting, but panned as being devoid of scientific reality, I could take some solace in knowing that at least some scenes that I touched, were both fun and rang authentic.

Among the many movies I worked on were "A Beautiful Mind" (2001) starring Russell Crowe, the first "Lord Of The Rings", and many others. However, one goal I had set for myself early on was to do something – anything – on some sort of Star Trek production. That dream had come true two years earlier, in 1998, when I got my chance and landed an effects stint on "Star Trek 9: Insurrection".

Amazingly, I was the only one who was on the extensive digital visual effects crew at Blue Sky VIFX who was also a "Trekkie" (hardcore Star Trek fan, to the uninitiated). This proved very useful because, for example, the CG artists doing the phaser blast effects were using the wrong colors with the wrong weapons. This and other things I noticed earned me extra duties as "Trek consultant" on top of my already busy visual effects schedule. Visual effects supervisor Jim Rygiel made good use my Trekkie skills to make sure scenes were being put together authentically and consistently with the rest of the Star Trek universe. One of my rewards for this extra work was an afternoon of playing with some of the rare props from the movie – every Trekkie's dream!

Close Encounters Of The Hollywood Kind
All the while I had my sights set on one thing: I wanted to be on staff as an artist with Walt Disney Studios. If I could pull it off it would be the crown jewel of my career push at that time. With that in mind, in 2000 I reached a critically interesting intersect. As "Hollywood" as I had gone, it was my science background that still continued to shine through. Mars has a resurgence of interest in it, as it often does in a seeming on-again off-again cycle. In this case, two studios simultaneously vied for my services on two competing Mars films. On one hand was Digital Domain who wanted me to work with them on "Red Planet" – back then considered the more "serious" film. The other was Disney Studios for "Mission To Mars". The only caveat with Disney was that it would be with a sub unit known as DreamQuest Images – an independent company located in Simi Valley, but with ties to Disney. I banked that the DreamQuest job on "Mission To Mars" might somehow land me directly at Disney Studios – so that was the job I accepted. The folks at Digital Domain took my choice as a sort of affront and vowed that they would someday get me on one of their films (which eventually they later did).

Interestingly, in almost prophetic fashion, I learned when I joined the "Mission To Mars" team that this film was actually focused on aliens – seemed I could not shake off the whole UFO thing – even more than I realized. The killer for me was the fact that the climax of the movie was actually focused on – unbelievably for me – the Face On Mars! Yes, I was actually working on a movie where the very thing I had spent years debunking while at NASA was being portrayed in a major motion picture, as a legitimate ancient alien artifact. The irony was not lost on my NASA buddies, who continually ribbed me about this. Seemed that life was steering me into the UFO world – slowly – but definitely – regardless of whether I knew it or not.

As I settled in, I found out that a fellow ex-JPL colleague of mine, Matt Golombek – also a planetary scientist, was "technical consultant" on "Mission To Mars", while I was an employee there working as a member of the visual effects team (and I also acted as "science advisor" on an as-needed basis). The studio was very pleased with having two scientists on board their staff. And for me, having a fellow NASA person on board this production made the whole Face-On-Mars thing a little easier to bear – and in the end I chalked it all up to just good imaginative fun. Cydonia and the Face On Mars faded from my focus, as I had fun bringing various effects to life. Not only that, but the spacecraft the Disney imagineers had created for this film was, simply put, really-really cool! It really made the movie feel real and come to life. I soon came to realize that I had indeed made the right choice choosing to work on "Mission To Mars" over "Red Planet". The work was great, my co-workers were simply awesome people! And me? I was having the time of my life!

Meanwhile, as Matt typically dealt with a lot of the big-picture science of the movie, I stepped in only when it came to resolving issues revolving around specific visual effects. Although I made lots of nice contributions here, it's the one so-called battle I lost that stands out most in my mind. It escalated all the way up the chain of command on the production, bypassed Matt (luckily for him) and landed squarely on the desk of the effects producer, story editors, and Ultimately, director Brian De Palma.

My beef? A scene I was working on dealt with M&M candies in space! Sounds fun, but here's the situation. For those of you who may not have seen this film, there is a notable scene where the astronauts traveling to Mars have a "wow" moment when M&Ms spill out & float around the spacecraft cabin. As they float around they begin spiraling and mimic a helix, thus causing the astronaut to think about DNA (because the DNA molecule, as we all know, is a double helix). The scene was critical to the astronaut solving a problem that was key to the plot moving forward. The M&Ms were CGI, and as I worked on the scene I couldn't shake the feeling that something was very-very "wrong" about it. Something fundamental. And then it hit me! It was indeed something *fundamental* – the fundamental laws of motion. Without getting into the weeds of physics, basically an object set in motion will continue that motion. How does that apply? It means that if the astronaut in the scene spills M&Ms, the candies will move in straight lines along paths of motion into which they were set. In other words the candies should be moving in random straight lines, not in spirals or circles. That would be impossible. Ultimately, for the sake of "story", my protest, though judged completely correct, was over-ruled and I was told to do the effects on the scene "as written" – physics be-damned. The main director's position: it was a pivotal plot point, and changing or dropping it could cause potential large story re-writes and re-shoots of certain scenes. Nope – the M&Ms were gonna be spilled and they were going to float around in helical spirals, and I was going to help make that happen – and that's that. That's what they deemed necessary, so having made my best case and losing, I set about "making-it-happen" with the fluidity and professionalism that had been my hallmark. The good news for me was that my ability to rapidly churn out high quality visual effects, and also act as a science advisor on the side and still be a "team player" even if it occasionally meant going against the basic physics, shined a very positive light on me.

This also meant I would be among a select group of freelancers who would survive to become fulltime regular staff at the prestigious Feature Animation division of Walt Disney Studios. How so? As it turns out, during production of "Mission To Mars", DreamQuest was in talks with Disney to be formally bought out and brought into the fold as an official part of the Disney Studios family. Unfortunately, negotiations soured as Disney wanted potential staff to lose seniority when transitioning from DreamQuest to Disney, thus saving Disney money, and hurting many career digital artists with years of experience and force them to, in essence, re-start their careers as newbies, and thus accept the low pay that comes with being labeled an entry level employee.

This unfortunate turn of events caused our animation union (IATSE Local #879) to step in, in our defense, and try to halt such an unfair action and negotiate a fair and equitable contract for us. As production wound down and freelancers began dropping out, those of us who "made the cut" (to become staff employees at Disney under this buy out) went into meetings with our union. Thus preparations were made to potentially go on strike to force a fair resolution. Long story short, ultimately, a resolution was reached. We kept our seniority, benefits and pay - and before I knew it I was working at Disney Studios Feature Animation in Burbank as a fulltime staff employee – a very prestigious achievement. It was one of the best jobs I ever had. Being part of the Disney family, to me, was a dream come true!

Two of my biggest dreams were to be part of NASA, and the other was to be an artist at Disney. I was flying high! I poured myself into my work, often putting in over 80 hours a week – and loving it! I would often shoot down the 5-freeway just to have dinner at Disneyland, hit a few attractions their, and then go back to work and eagerly put in a late night – no sweat and always a smile on my face. Disney treated me extremely well, and because of my extreme work ethic, I quickly found myself working simultaneously on five productions – and loving every minute of it!

By far, one of the coolest things that happened as a result of my hard work was being asked to work on the various famous and iconic historic spaghetti-kiss scene from "Lady & The Tramp". How was this possible? One of the productions I was on was "102 Dalmatians". In one scene in the film the animals watch a TV and the classic "Lady & The Tramp" is on. Unfortunately, (or fortunately for me), that scene suffered from some historically oriented defects. Though they may be forgivable in the context of the full

1952 film, as part of this new film, it didn't mess well. It needed to look fresh and match the look of the new film, but without ruining or significantly changing the original footage. Of particular interest was dampening down the appearance of "Newton's Rings" – faint circular-patterns that appeared here-and-there throughout the scene and were a product of Walt Disney's original multi-plane camera technique in that original film.

The fallout from doing such work was that the artist lucky enough to not just touch such famous footage, was that he or she would then also be able to include it on their demo reel (digital portfolio). That artist turned out to be me – and it was an honor to have worked on something like this. It was one of many choice assignments that I was blessed to have been part of.

However, as wonderful as things were, there was still the occasional drama. In my case, it turned out to be UFO-related. It's probably best I not detail the exact production or specific people involved, but the scenario was a brutally simple one.

I was working on a set of scenes and was handed some new footage to digitally edit and clean up. It was from the U.S. Air Force, and for now it's simply enough to say that it showed a successful historic jet test flight. My problem with it? A UFO was plainly evident in the shot. The footage was black and white and the object in question was a bright white orb that appeared in the distance, rapidly approached the aircraft, paced it, and then took off in a burst of speed. It was incredible. I kept playing it on my machine again and again. The powerful imaging software at my disposal allowed me to view the shot again and again in various ways – and the result was always the same: the UFO was actually "there" and very real.

Unfortunately, the UFO being real complicated my job tremendously. What was I to do? Clean up the shot and leave the UFO there? Were the Visual Effects Supervisor, Producer and Directors aware of what was in the scene? I tabled the shot and shot out a series of emails and made some calls and left some messages, and then went on working on other shots while I awaited an answer. What had I asked? Simply put I asked everyone if they were aware that there was a UFO in the film footage, and was that intentional – or if not, was I supposed to still leave it in, or was I supposed to "do something" about it? In essence my question was: "do you want me to digitally paint out and remove the UFO, or leave it in?"

Among people in my visual effects team word spread that I had a picture of a UFO and people were dropping by my office cube to see it. They wondered what I would do, and my answer was that I didn't care as long as management made a decision one way or the other. In hindsight I obviously cared, or I would not have been so adamant for someone else to make a decision. One of the effects producers, known to be somewhat of a hot head, dropped by and when I showed him the UFO he got visibly angry and demanded I simply paint it out. I told him that I would have no problem doing so as long as he took responsibility for the "cover-up". This joke, however made him uncomfortable enough that he huffed and puffed in anger and just stormed off. In the end, I got a call from one of the big shot producers who asked why that specific shot was held up. I explained the problem and he countered that he couldn't understand how something like this could have happened since they had acquired the footage directly from the Air Force. When I asked about the details, he refused to provide the details, and voicing exasperation, said to not do anything until he had a chance to see it.

In the end I was told to paint out the UFO as if it was never there in the first place. Word was, from my friends that some sort of Air Force official who personally brought the footage was very upset with the whole affair, but pleased that I had agreed to paint the UFO out without question. While I worked to comply, word spread among our team that I had UFO footage and people dropped by to take a peak before it was gone. Everyone seemed keenly aware that their jobs could be threatened, so they quietly satisfied their curiosity and then went back to their respective jobs. At my desk they all oo'd and awed, but outside the studio no one ever said a word. As I finished work on that project, I deleted the original footage. Only my painted version with no UFO remained. No one ever spoke of it again – accept for a few - in hushed tones. After a while even that ended.

In essence I had, without even realizing it, now become part of the UFO cover-up conspiracy. I had no choice. I had done what I had to do to keep my job – and moved on…

Postscript - From The South Pole To Mars

As I moved from job to job in Hollywood I worked on a number of absolutely amazing projects and met some wonderful, talented and very nice people. The people in animation and visual effects are truly good-good people. In the meantime my father had gotten cancer and my mom was completely overwhelmed with trying to take care of him. Even though I very much enjoyed working in Hollywood, it did require endless hours. My workweeks often topped 100 hours, and although I didn't mind, it made it extremely difficult to help my parents.

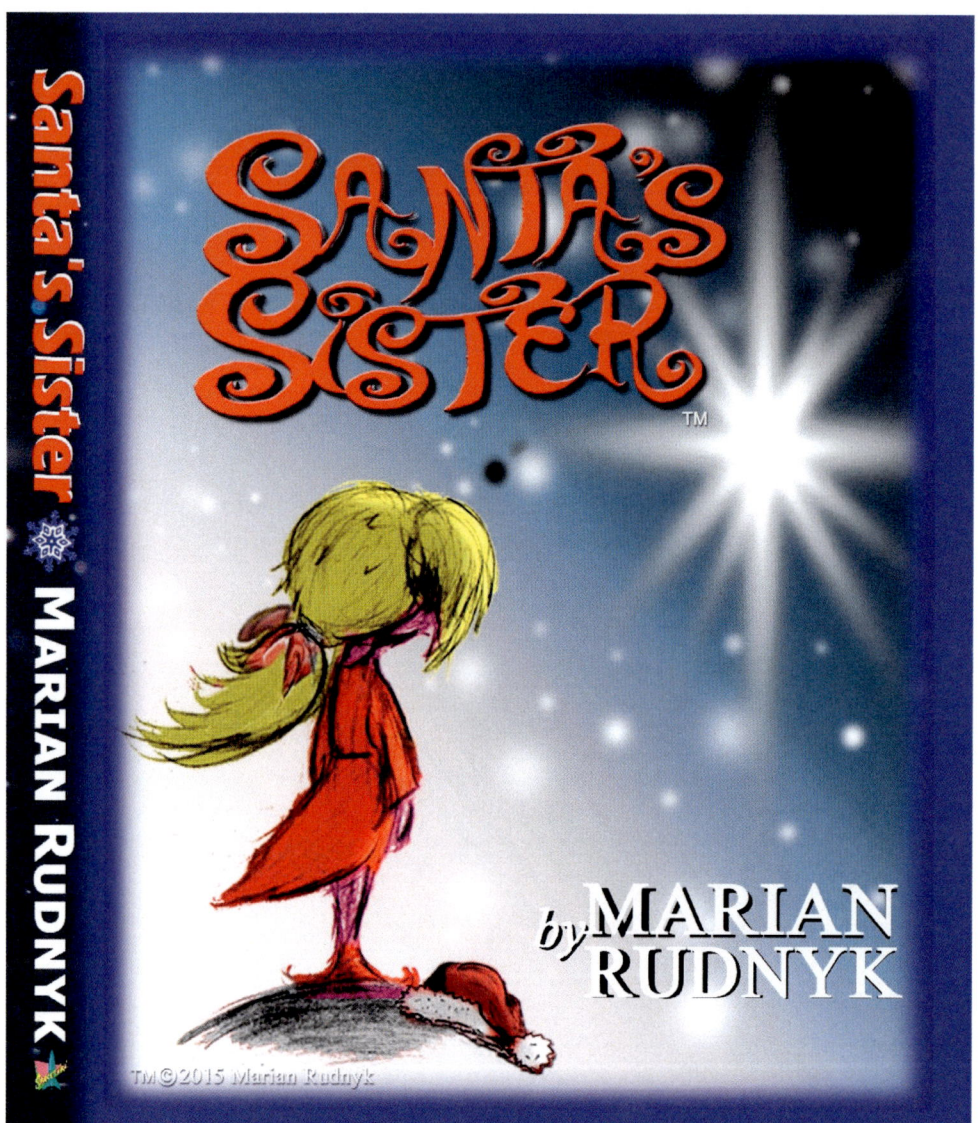

The cover of the paperback edition of Santa's Sister, released via Amazon in December 2015 as an e-book, and in a deluxe large-sized paperback form in 2017. It is the story of a little girl who believes Santa Claus has a sister named Noel (seen on the book cover) who lives at the South Pole and decides to go and find her. It's an epic family-friendly grownup fantasy story set in Christmas... and even has a NASA twist. (©2015 Marian Rudnyk)

In the end, I decided to move on to address one of my other many passions: writing! I had a number of books I had been working on, on the side. I could work from home in pursuit of this new career path and thus at the same time help my parents. It would be a win-win. Out of the over 64 writing projects I had started, I chose an epic Christmas story called "Santa's Sister" as my first project to launch me in my new direction.

However, Fate had one last stab at me, as in one last twist of irony I worked on my last movie project (before becoming an author), at a studio called CIS Hollywood. While there I worked with none other than Ken Jones – an old friend from JPL who was formerly part of the 1970's Viking Lander team that landed on… Mars!

Ken Jones (in the Viking helmet) & Steve Wall, just hours after the landing of Viking 1 Lander on the surface of Mars. (Figure 21, page 26, "The Martian Landscape", NASA SP-425, 1978; courtesy NASA)

Ken Jones, definitely one of the more colorful imaging scientists on the Viking Lander Team, lives up to expectations (he's the one with the Viking helmet in the photo at left) in this great picture that was featured in NASA's official Viking Lander book (that was released at that time), called "The Martian Landscape". To this day it remains one of the premiere go-to resources of information on this critical Mars mission. It offers rare detailed insider glimpses of Mankind's first successful landing on Mars as well as the people behind it. It even includes engineering information as well as stereo pair pictures – with 3D glasses included! Ken was a member of the team that reconstructed the first pictures radioed to Earth from the surface of Mars by the Viking Lander. To this day I still wonder what he ever did with that awesome Viking helmet…

Seeing Ken again, in a Hollywood setting, was both weird and a very pleasant surprise. If that wasn't odd enough, my stint there at CIS Studios was for one of my last movies I was to work on, "Scary Movie 3" which featured… UFOs and aliens in the story - and those were the scenes I was assigned to…!

2:

The Main Event – A Firsthand Account

Some quick thoughts
Although the "Hollywood UFO incident" (as I like to call it) had, in a weird twisted way, quietly become legend among my close NASA pals, I rarely spoke of it to anyone else at all. And as far as UFOs went, it all faded from consciousness. In the meantime my brother, a longtime "UFO fan", became more actively involved in the actual UFO community and their world. In the process he somehow blended his newly found formal credentials in philosophy with UFOlogy. For my part, I either pushed all discussions, of UFOs and aliens aside, or simply shrugged off his (and other people's) comments as silly and not worthy of my time. That I had essentially covered up UFO footage was beside the point to me. I compartmentalized it away as a one-off oddity and mentally filed it away under the "*whatever!*" file. I also used the fact that that footage was old, as a rationalization to deride it as having no bearing on the current world's affairs. But the fact of the matter was, I would soon be forced to reassess everything I knew, and in ways I never expected…

And Then It Happened
As you recall from the start of this book, I was at a McDonald's, walked outside, and then saw four UFOs. But it wasn't quite that simple. So what exactly did actually occur?

Lots of people supposedly see UFOs – why was this different? Why was I different? Why would my encounter matter to all the government people who lived in the shadows? The answers are not easily evident. One thing's for sure: I greatly underestimated everyone's interest – and suddenly writing this book became almost like a moral imperative. An important story that needed to be told – while it could still be told. And I can't stress this last point enough. This book was not supposed to come out. This story was not supposed to be told. But the "power that be" have brought it on themselves – so here we go!

To understand how my sighting happened, it's important to realize that a number of odd seemingly unrelated things sort of "aligned", if you will, to set the stage for my sighting. Sort of the perfect UFO-storm, if you will. It was a combination of: calendar timing, the weather, a gift, my car and a diner – then the aliens and the government did the rest. Now, let's plunge headlong into my strange journey…

Setting The Stage
So let's go back to New Years' Eve of 2016 (December 31, 2016), to a little suburban town in Southern California, called Monrovia – just North-East of Los Angeles, in the Pasadena area – nestled among the green rugged foothills of the towering San Gabriel Mountains. Founded during the Victorian era of the late 1800's, President Teddy Roosevelt had made a historic speech here, and many streets are lined with well-preserved historic Craftsman and Victorian homes. The "Main Street USA" character of the town has made it the continuing scene of countless Hollywood movies and TV shows.

It had been a very cold and rainy New Year's Eve. The next day, New Year's Day (January 1, 2017), was no different – not very typical New Year's weather for the Los Angeles area, that's for sure. That being said, one thing that was also atypically unique about that day: that was its day of the week. This may seem like an off observation to make – but it's actually very-very important. It goes something like this: New Year's Day was on a Sunday, and for Southern California, the so-called "rules" dictated that the usual massive New Year's Day celebrations (mainly the Rose Parade in the morning, and the Rose Bowl football game in the afternoon) were all delayed until Monday, January 2nd. This is critical in understanding the contextual setting of my sighting.

This meant that after an evening of partying for New Year's Eve (on Saturday night), on Sunday morning (with all the festivities bumped to the next day, on Monday) that everything was not just quiet – but was literally *dead*. Everyone was simply recovering from the night before – with nothing to do until the next day after. And with the weather uncharacteristically cold and wet, the traffic was beyond light. Thus, barely anyone was out or about. The town, and surrounding areas were literally dead. Overcast and cold, most people simply stayed in to rest in preparation for the delayed festivities to be held on Monday.

Additionally, I had recently completed a family book project for a friend, and as a bonus she had given me an old spare digital camera she had won in a raffle a while ago, and didn't need. So we now have the weather, the calendar day, and the gift all lined up. The so-called stars of fate were beginning to align. Next came the car and diner…

I had spent New Year's Eve with my mom, so that she wouldn't be alone. We had a nice time. Then on New Year's Day we decided to at least do *something* to celebrate, even though the whole town was dead – bad weather be damned. It was a no-brainer: we both loved the local McDonald's because it was one of the last remaining "McDiners".

The retro décor, vintage music, friendly staff, and stunning mountain-view wrap around windows made it a natural attraction to not just the Route 66 sect, but to many locals who became regulars there. This gave the McDiner a rare cozy feeling – a gathering place with friends, family, good music in a unique warm setting. So deciding to go to the McDiner that day was a no-brainer - we would have a late lunch there.

So you can appreciate this cute little diner, here are two pretty nighttime pictures I took of it in June 25, 2016 – by chance, only five months before my UFO sighting.

The McDonald's McDiner on the evening of June 25, 2016. The top image (on previous page) shows the building's North-facing side (which faces the local San Gabriel mountains, which is where the UFOs were heading). Notice the nice neon that envelopes this beautiful retro-styled building, and the gorgeous glass brick corner wall at the entrance (right side under the vintage "Speedee" neon character sign). The bottom image (above) shows the West-facing side of the building. You'll notice my 1962 Ford Thunderbird on the left side of that image. I own several classic cars, and taking pictures of them at locations such as these, is one of the things I love to do. Unfortunately, the McDiner no longer looks like this. Its charm was destroyed by a destructive renovation in October of 2017, oddly, shortly after my sighting. (©2016 Marian Rudnyk)

Once we headed out we got into my classic 1967 Pontiac Catalina convertible. I had a brainstorm! I told my mom to wait a minute as I ran back into the house and grabbed my (gift) camera. Why? I explained to my mom, as I got back inside the car and we headed out, that with everything sort of "dead", there would probably be no one at the McDiner, and therefore I could get some nice pictures of my car alone next to the cool-looking retro diner. It was an incredibly fortuitous decision, because it would mean that I would have a good digital camera with me when the UFOs arrived. Now all the pieces were in place.

As we pulled into the McDiner parking lot and parked, it was about 2:45pm. Asking my mom to wait inside the car for a minute, I got out and, starting at 2:47pm I quickly shot

15 pictures of my car and the McDiner from various angles. At about 2:50pm I shot the 16th picture – it was of my mom in my car. And with that done, I put away my camera and helped my mom (she was 76 at the time, and struggles with multiple health conditions to this day) to get inside the McDiner, picked a both and sat down. With no other customers there we had our choice of the best booth so we picked one with the best scenic outside views.

At 2:51pm I ordered our food. How do I know all these times so specifically? You can look at the photos in "Apendix-1: Photo Archive". All the pertinent photo info is shown there in its entirety, along with an actual thumbnail of each picture. Additionally, "Appendix-4: Support Materials" includes a scan of our actual McDiner food receipt, which has a date-time stamp on it.

So, now back to the matter at hand: our lunch. At 2:58pm I took two more pictures (one of a McDonald's Christmas display (was still up), and one of my mom sitting at our booth, waiting for our food. Here is that photo. I'll explain why I show it here in a sec.

Here is Photo #1674, taken by me at 2:58:16pm (the day of my sighting) of my mom at our booth. Notice all the windows and how they offer spectacular views in nearly all directions. (Image #1674 taken at 2:58:16pm with Coolpix S3100 camera ©2017 Marian Rudnyk) The worker then brought our food out for us (the McDiner used to offer partial table service). As I sat down I asked the server to take a picture for me of my mom and I together in our booth.

Here is Photo #1676 (taken at 2:58:44pm) of my mom and me, happy to be sharing our newly arrived McDiner meals on New Year's Day 2017. Notice the heavy coats – a reflection of the cold rainy weather outside. (Image #1676 taken at 2:58:16pm with Coolpix S3100 camera ©2017 Marian Rudnyk)

It was all innocent enough. Nothing out of the ordinary. As you can see by how we were dressed it was obviously a very cold day. Little did either of us know anything extraordinary was about to happen.

Before I continue, I want to note how lucky I was to have been at this particular place, and at that time. You see, one of the things that was also special about the McDiner, was its fish-bowl-like quality. When you sat inside it to eat, nearly anywhere you sat you were practically enveloped in stunning panoramic views of the outside. What a treat! And also critical to noticing anything "out-of-the-ordinary" that might be "flying around"…

To the North you could see the San Gabriel Mountains in their near-complete entirety. You also could see to the East. And to the West you had a full view of the horizon from South to North (left to right, when you faced West). It was a spectacular viewing location. Unfortunately, as I noted under one of the previous pictures, the McDiner doesn't look like this any more. As if almost on purpose, it was destructively renovated shortly after my sighting. It's now dark and cluttered and most of the views to the outside are obstructed one way or the other. Even the stunning glass brickwork is gone. Fortunately, I did take a few pictures of the inside. They are not perfect but they do at least somewhat highlight the amazing fish-bowl panoramic views from inside and the clean unobstructed open airy feel of the diner.

These images are from June 24, 2017(left, only six months after my sighting) and Sept. 11, 2017 (right). In the left picture you can see the view from the same booth we sat in on January 1st, looking West and South-West. In the right picture you can see that massive expanse of windows that faced North, but also gave clear views to the West as well, especially if you sat where my mom is sitting in that picture. On the day of the sighting, I was sitting where she is in this picture. A couple weeks later, by September 25th, as I've previously mentioned, this would all be gone. (Left image #DSCN2417 taken Saturday, June 24, 2017, 7:24:14 AM; right image #DSCN2810 taken Monday, September 11, 2017, 4:29:10 PM; both images ©2017 Marian Rudnyk)

This stunning picture taken Sept. 10, 2017 of the reflections in some of the west-facing windows show the wonderful panoramic views that the original McDiner provided. (Image #IMG-20170910-174147-1CS taken Sunday, September 10, 2017, 6:41:48 PM with BLU Studio-G smartphone; ©2017 Marian Rudnyk)

The Flying Saucers Arrive!
As we ate and chatted, the time passed. As is normal when eating at the McDiner, our eyes continually scanned the panorama that surrounded us as we talked and ate. At about 4:22pm I looked up and noticed four objects suddenly drop out of the clouds in the Southwest. They were in a diamond formation. Their decent was very smooth and quick. Then they slowed and started moving directly North. I mentioned to my mom that some sort of aircraft are flying outside. I instantly assumed that they must be part of the New Year's Day celebrations. But just as quickly as we had made that observation, we commented that that was impossible because everything was delayed until tomorrow (for the reasons I've previously outlined).

Now my curiosity was peaked. Barely a few seconds had gone by and the objects were still easily in view. I told my mom I was going outside to take a closer look at what exactly they are, and rushed outside. What I expected to see, I thought, were probably some sort of parade aircraft or perhaps manned parade balloons of some kind. Perhaps, I figured, they were rehearsing for tomorrow or something.

However, nothing would prepare me for what I actually saw.

I stepped outside of the McDiner, and looked up. What I saw stopped me dead in my tracks. As I stood there I watched, not one – not two – not three, but four disk-shaped craft silently move below the low cloud deck. My eyes widened with the realization: I was watching four flying disks – what were commonly call flying saucers. They were real. You would think that at such a revelation I would get emotional, either freaked out or excited. Not me. Instead, my mind hyper-focused! I snapped into science-mode – clear minded and calculating. There was no time for emotion. First, I decided to assess what I saw: the appearance, movement and any other distinguishing features.

In my head I quickly started to note as many specifics as I could. They were moving from my left to right (directly South to North) in the western sky below the cloud deck. They appeared, at first, as dark and very distinct disk shaped silhouettes against the thick cloud deck. The air was crisp and cold, with gusts that caused the flag at the McDiner, and the tall massive oversized one at the Living Spaces furniture store, to both flutter to the East. I could hear the clanging of the rope and chain of the McDiner flag clanging against the pole – almost hauntingly given that traffic was light – and that the four craft were flying silently. Their movement was very smooth and determined, and they headed directly from South to North, perpendicular to the West-to-East strong wind gusts. I was particularly struck by how they were completely unaffected by the winds – especially given that the gusts were very strong. They made no sound of any kind. I now winced, trying to see details and could see absolutely no flight surfaces of any kind. No wings. No engines. No rudders. No propellers. No nothing. Just very-very distinct disk shapes. But it was their movements that especially caught my attention. The last disk was trailing and dropping quickly, as it struggled to move forward, and the others seemed to be reacting to this – their formation was coming apart – as if they were somehow trying to help it.

Although these assessments took a bit to write up here, in reality, in real time, I made them very quickly. Even so, I was suddenly hit by a sudden realization, and thought to myself: "What the heck am I doing!? I have a camera with me!" And not a smartphone with a camera, but a full-on regular digital camera. This was extremely good news. The bad news was that it had been a recent gift, and although I had gotten familiar with it, I was still getting a "feel" for it.

There was no time to be wowed or awed. I had already made my observational assessment and I now immediately realized that I needed to document what I was seeing. I wanted to be able to not only prove what I saw, but to also study it later. I needed data: pictures and video. So now, the scientist inside me kicked into overdrive, and I began to calmly and methodically photograph what I was seeing. I pulled out the camera and got to work documenting what I saw – and what I saw was amazing…

As I set myself to take my first picture I noticed that the disks were breaking their formation even more now. That last disk was still dropping, and beginning to trail and lag behind the others. Their formation was breaking apart and they were starting to spread apart. At that point I took my first picture. I made it a full view shot, so that I could capture the surroundings at that moment too, and have context for future shots. Here is that first picture I took…

Here is the raw original version of my first UFO picture, Photo #1677, taken at 4:22:58pm. Note that the time-stamp on the picture (in red on the right), is off by one hour because the camera was not set for Daylight Savings time. This is true for all pictures I took that day. (Image #1677 taken at 4:22:58pm with Coolpix S3100 camera; ©2017 Marian Rudnyk)

And below is the same picture, but now with the craft marked in so that you can see better what I saw. Because cameras make things "appear" smaller, the picture doesn't convey the sense of "presence" that these craft had. For this reason, I will present each image this way (raw, then marked, and then with full labeled version that includes enlargements) so that you can get the full impact of one.

Here again is Photo #1677, taken at 4:22:58pm, now marked to highlight the position of each craft. (Image #1677 taken at 4:22:58pm with Coolpix S3100 camera; ©2017 Marian Rudnyk)

By this time, it was easily clear to me that these were not so-called "objects". A chair is an object. A rock falling through the sky is an object. These, however, it was very evident to me, were under intelligent control, and were being "flown". Their movement was deliberate and controlled. They moved intelligently, and with intent, in a formation - so they were obviously piloted (either from within, or remotely). So from this point on I will be referring to them as craft. As time progresses, during this sighting, and beyond, and as my understanding of them grows, the terms I use to describe them will evolve too. At this point I had, at the very minimum, a definitely defined sense that these were indeed some sort of "craft".

Although these first pictures look like fuzzy spots – trust me – they were clearly and easily visible to the naked eye. By the time we get to the final pictures, you will be as shocked as I was when I finally got to view the pictures on my laptop, because they clearly showed flying disks – with details and all! That being said, as far as the picture naming goes, I have kept my labeling simple, and assigned each craft a number, one through four, with the lead craft in direction of flight being "Object/Craft-1". For now I will use the words "Object" and "Craft" interchangeably (we will address the issue of naming later). This means that Object-1 is the furthest on the right, and the last craft (the one dropping and trailing, is "Object-4"). This is the numbering I will maintain from here on out. Now let's continue with a first look at the pictures in their raw form.

To stay chronological, the next image is from the video. As the craft continued moving North, I switched my camera to video mode and shot 34 seconds of video to capture their motion. Unfortunately, the camera had a hard time focusing the craft against clouds and so the craft came out a bit fuzzy. When viewing the video itself this is evident because the craft appear to strobe in-and-out of focus. Because the camera was new to me, I didn't know how to adjust for this at the time, and could see the digital tracking squares struggling to lock, but I just shot the video as-was, in the hope that it was better than having no video at all. Later experiments with the camera showed that it had no trouble focusing on, for example, airplanes against clouds, so it has since become clear to me that there was something possibly in the nature of these craft that made the video tracking function "struggle" to lock-in properly. However, fortunately, the camera did well in picture mode. For the purposes of this book I have made a series of screen-capture frames directly from the video. These are included in the back of this book in *"Appendix 2: Video Screenshot Gallery" (I will also be making the full video available online).* Below is a sample of two of these screen capture frames from the video…

This frame capture is from the very beginning of the video. As can be seen, all 4 craft have moved to the right (north) from where they were in the first photo. The resolution of the video is not as good as that of the photos, but the objects are significant enough that they still show up. (Screencap from video #1678 taken Sunday, January 01, 2017, 4:23:40 PM ©2017 Marian Rudnyk)

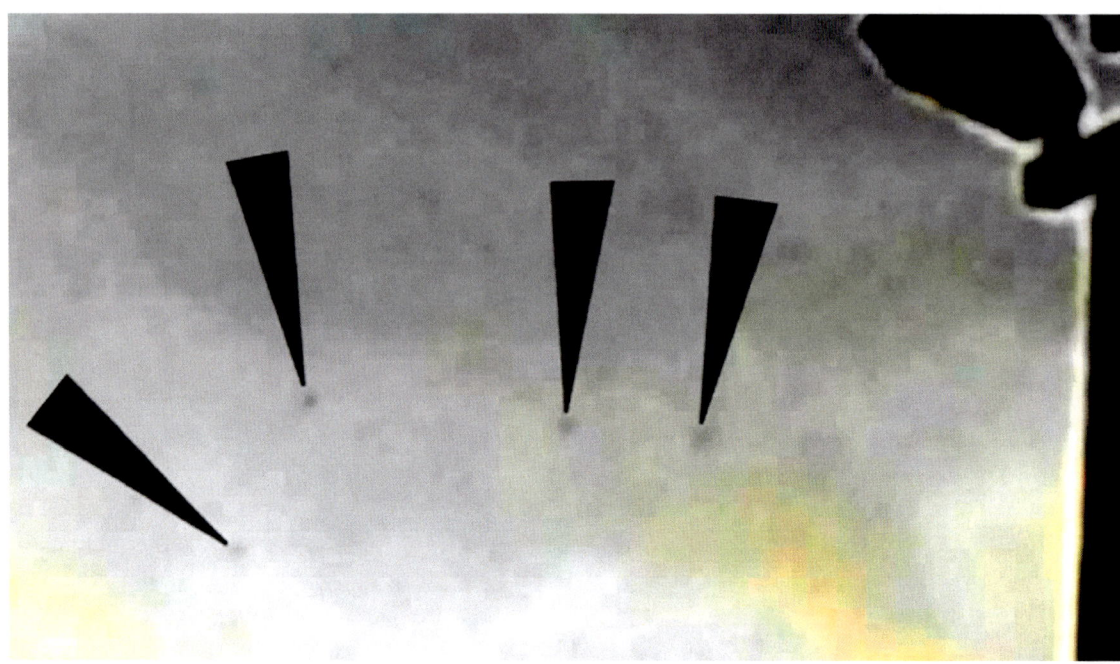

Frame "T=4" (which is ~4 seconds in), from my video (#1678.avi). This particular frame is a close-up, and has had the contrast enhanced to make the craft easier to see. Although there are splotches in this image, the craft are easily identifiable because of their motion from frame-cap to frame-cap. The large dark object on the far right is a lamppost in the McDiner parking lot. When taking the pictures and the video, I made sure to try to include surrounding objects that would provide reference context. (Screencap from video #1678 taken Sunday, January 01, 2017, 4:23:40 PM ©2017 Marian Rudnyk)

Stopping the video and letting it "save", I then changed the camera back to picture mode, and carefully zoomed in. I shot three more pictures of the craft. On a computer screen they are very easy to see, but for the purposes of this book I am including a "marked" version of each frame, for now, so as to make them easier to see (we'll have a closer and much more dramatic look later). Below are those original images…

Raw version of picture #1679, my second "UFO picture", taken at 4:23:52pm. (Image #1679 taken at 4:23:52pm with Coolpix S3100 camera; ©2017 Marian Rudnyk)

Marked version of picture #1679, taken at 4:23:52pm. (Image #1679 taken at 4:23:52pm with Coolpix S3100 camera; ©2017 Marian Rudnyk)

As you can see here in #1679 (above), the craft were beginning to re-establish their formation. In this frame, Craft-4 was at its lowest point, and thus closest to me. I could actually make out panels on its underside! It appeared to still be struggling, but the sense I got from watching them is that somehow, from the way the other three craft were re-positioning themselves, they were somehow managing to somehow pull Craft-4, through some unseen force, back into formation and thus help it along. (If they failed, then Craft-4 would have literally dropped into the middle of town!)

Raw version of picture #1680, my third photo, taken at 4:24:00pm. The objects are very prominent in the sky. (Image #1680 taken at 4:24:00pm with Coolpix S3100 camera; ©2017 Marian Rudnyk)

Marked version of picture #1680, taken at 4:24:00pm. (Image #1680 taken at 4:24:00pm with Coolpix S3100 camera; ©2017 Marian Rudnyk)

Here in frame #1680 (above), you see further evidence of the craft re-establishing their formation and actually helping craft #4 fly with them. They were all now gaining altitude. Additionally, my camera was frustratingly slow, taking nearly 10 seconds at times just to reset for the ability to take the next picture. As it did, I quickly pressed the shutter button and got my last so-called "UFO picture" (below) at 12 seconds later.

Raw version of picture #1681, my last UFO picture of the day, taken at 4:24:12pm. (Image #1681 taken at 4:24:12pm with Coolpix S3100 camera; ©2017 Marian Rudnyk)

As you can now see in this last picture, #1681 (raw & marked, above & below), the first 3 craft have helped Craft-4, and the four have now nearly re-established their original diamond formation, and were pulling away as they gained altitude. It's hard to describe exactly why or how I knew, but somehow from watching them, I got the very explicit sense that Craft-4 was *still* somehow in trouble - in distress - but its 3 pals had somehow managed to help it… for now.

Marked version of picture #1681, taken at 4:24:12pm. This is the last so-called "UFO" picture that I took that day. (Image #1681 taken at 4:24:12pm with Coolpix S3100 camera; ©2017 Marian Rudnyk)

After shooting those pictures I turned my attention to the area behind the craft, and the sunset. I didn't want to miss anything, and couldn't shake the feeling, for some reason, that there may be more craft, though I couldn't see them. So I shot two more pictures with different camera settings, in that direction.

Raw version of picture #1682, taken at 4:24:28pm. (Image #1682 taken at 4:24:28pm with Coolpix S3100 camera; ©2017 Marian Rudnyk)

Raw version of picture #1683, taken at 4:24:36pm. I changed the settings for this picture, which unfortunately, gave it an over-exposed "blown-out-sky" look. The yellow car in the foreground is my 1967 Pontiac Catalina convertible. This was the very last picture I took that day. (Image #1683 taken at 4:24:36pm with Coolpix S3100 camera; ©2017 Marian Rudnyk)

At this point, as you can imagine, I was eager to go back inside and tell my mom about what I saw and photographed. But I was torn. If I went in to tell her, the craft might be gone. So getting her to try to see them may be pointless. But at the same time I wanted her to see. I glanced up the craft. They were still there – moving silently towards the distant mountains. I decided to risk it and headed inside. As I opened the door I decided to glance back one more time – but they were now GONE! I looked around. They were nowhere to be seen. So that solved that dilemma. What struck me at that moment, as I looked around one last time, was that for the 4 craft to simply vanish, they had to have moved away in an immense burst of speed (in that second that I had turned away), in order to have either ducked into the distant clouds or mountains. This fact really made an impression on me as to the high level of technology that was displayed here.

I glanced back at the mountains and wondered: if Craft-4 was indeed struggling, might it actually wind up crashing somewhere in those mountains? The thought intrigued me. It was a tantalizing possibility. Already secretly disappointed that it hadn't come down mid-city, I secretly hoped it had dropped somewhere in the mountains – imagine the possibilities! In time, I would discover that I wasn't the only one who would entertain such thoughts, and many more…

What The Pictures & Video Revealed…
With the craft gone I rushed back inside and told my mom all about everything that happened and what I saw. I brought up the pictures in the camera, but the viewfinder was so tiny and grainy that it was tough to make anything out. I could see the disks – and there were details, but the poor screen quality coupled with a limited zoom ability frustrated any efforts at a real detailed analysis.

As we finished our desserts we contemplated what this could all mean if the pictures confirmed what I saw. Yes, if you're wondering, you did indeed read that right. Amazingly, even though I know what I saw, and had been a very methodical observer, I was actually trying to "self-debunk" myself and was leaning on the pictures and video I shot to prove to myself that what I saw was real. In hindsight it seems silly, but at the time, considering the enormity of what I saw, it made perfect sense. My whole world-view was being spun-180, so naturally I wanted to make sure beyond the shadow of any doubt that it was real.

The more my mom and I discussed the possible implications, the more we realized we need to get home – fast. We quickly finished and headed home. As I drove us home I made one quick detour and drove past a bank with a thermometer-clock display sign on Myrtle Ave. I wanted to note the temperature. It showed 62F. Also, Monrovia is home to many large and tall trees and very tall palm trees, so my mom and I were careful to note that the gusty winds were consistent and in the same west-to-east direction.
When I got home, which is nestled up in the foothills directly under the mountains, we noted the exact same conditions (wind and temperature). To the average person all this attention to detail may seem like overkill, but to a scientist like me, it was my way of locking down as many details as I could so that I could eliminate any stray possibilities and have a clear and true explanation for what I saw. And although I felt confident in what I saw, I knew that the pictures and video would be the most revealing. Would they confirm what I saw, or was there some weird explanation that I hadn't considered that would suddenly be revealed by them. We were only moments away from finding out.

Once inside at home I fired up my laptop. As it booted up I took the time to quickly sketch out what I saw. When I was done, the laptop was ready so I inserted the memory card from my camera into it. As I started to look through the imagery I was stunned. The reality exceeded my wildest expectations. One craft in one certain frame in particular simply jumped out at me. At that moment any and all doubt was erased from my mind. I now had spectacular confirmation. The so-called craft were indeed craft like no known earthly aircraft. All the frames I took consistently showed the same thing. The one spectacular from was #1680, in which zooming in that one trailing disk, Craft-4, proved to be extremely revealing.

Here, finally, is frame #1680, but now with enlargements of the 4 crafts in the image, montaged in, followed by a zoomed in close-up of Craft-4.

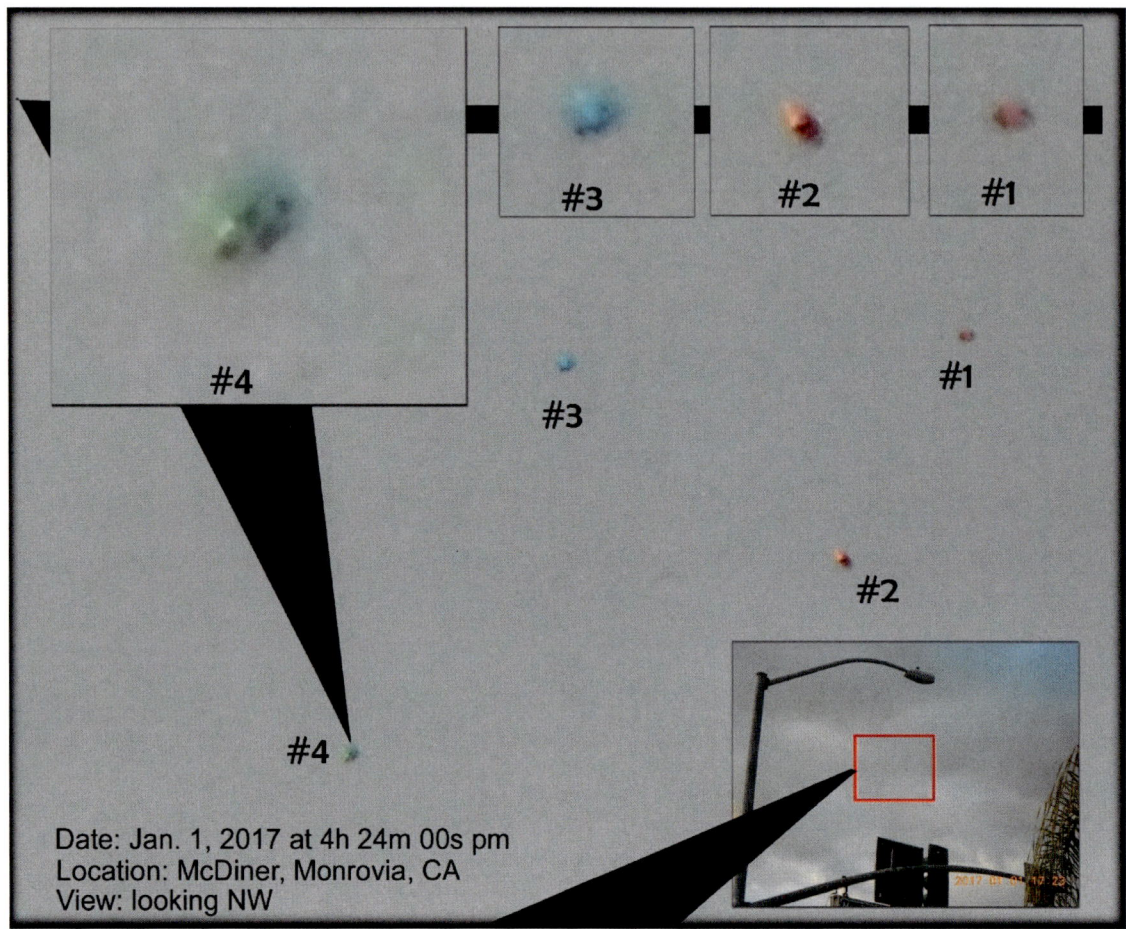

Frame #1680 close-up montage. Shown here is the original picture in the lower right hand side. The area marked by the red box is the area contained in the large main frame here itself, but now enlarged. Additionally, each of the 4 craft have been further zoomed in and shown in corresponding boxes. All consistently reveal the craft to be disks with triangular panels on their undersides. (Montage version of image #1680 taken at 4:24:00pm with Coolpix S3100 camera; ©2017 Marian Rudnyk)

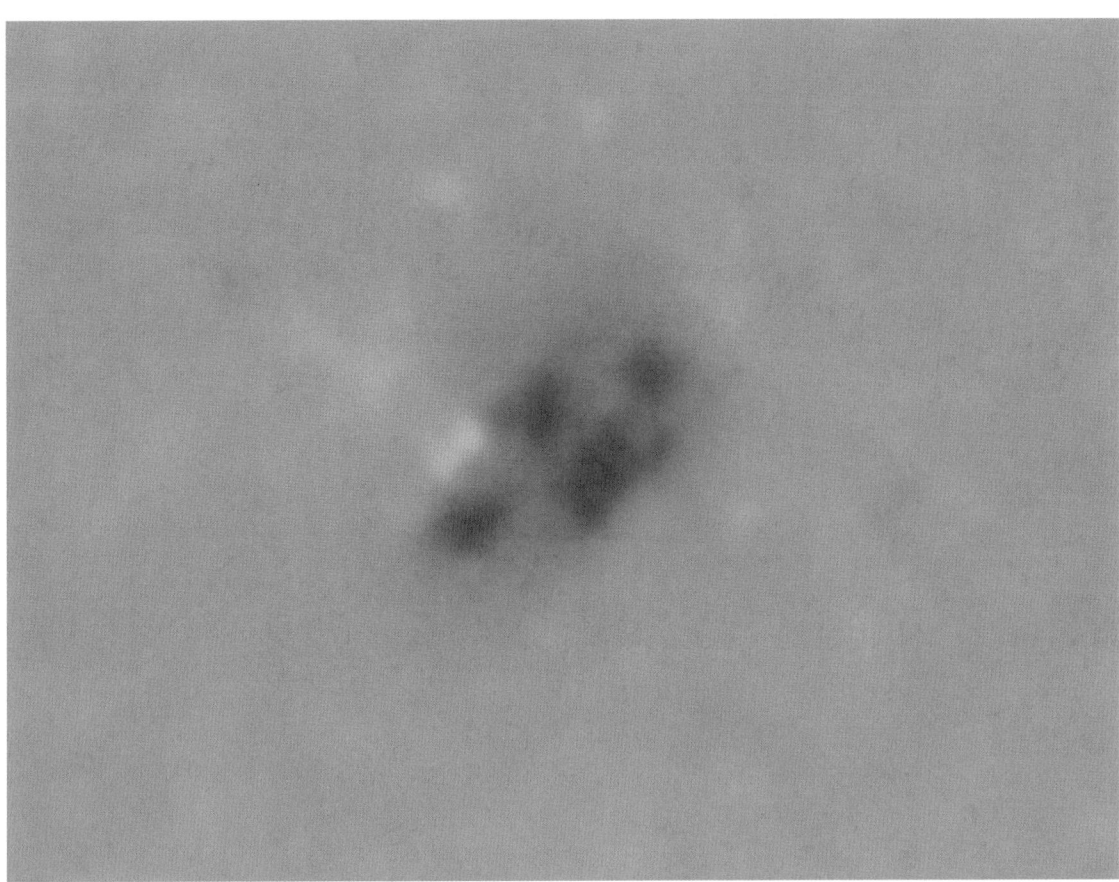

Craft-4, as seen in frame #1680, is so clear that it can easily be further enlarged. As seen here, then, is Craft-4, now visible in spectacular detail! Clearly visible are panels and a dark central area on its belly. (Close-up of Craft-4 from image #1680 taken at 4:24:00pm with Coolpix S3100 camera; ©2017 Marian Rudnyk)

And now below is the sketch I made of what the craft all looked like to me:

My rough pencil sketch showing details of the craft I saw. Notice the dark patchy area (on the disk's right side). In the original photo it appears as an area of splotches that don't seem to conform with the craft's design and symmetry. After further examination I now firmly believe this area is actually the damaged portion of this craft, possibly burn marks, or worse. We'll get into this a little further along, but I think it's important to note already here. This craft was behaving like it was damaged (probably by lightning), and it looks like the proof is here: Craft-4 was crippled. (Craft-4 rough pencil sketch; ©2017 Marian Rudnyk)

Before I continue, here are the remaining frames, with montaged in enlargements of the craft in each frame.

This again is the first frame, #1677, now labeled and with enlargements of the craft included. Because this picture was shot without any camera zoom, it shows the least amount of detail for the craft, but you can still clearly see that these craft are indeed disks and the barely visible, but there, dark splotches are the panels. Some of the disks also have a little bit of a different orientation. (Montage version of image #1677 taken at 4:22:58pm with Coolpix S3100 camera; ©2017 Marian Rudnyk)

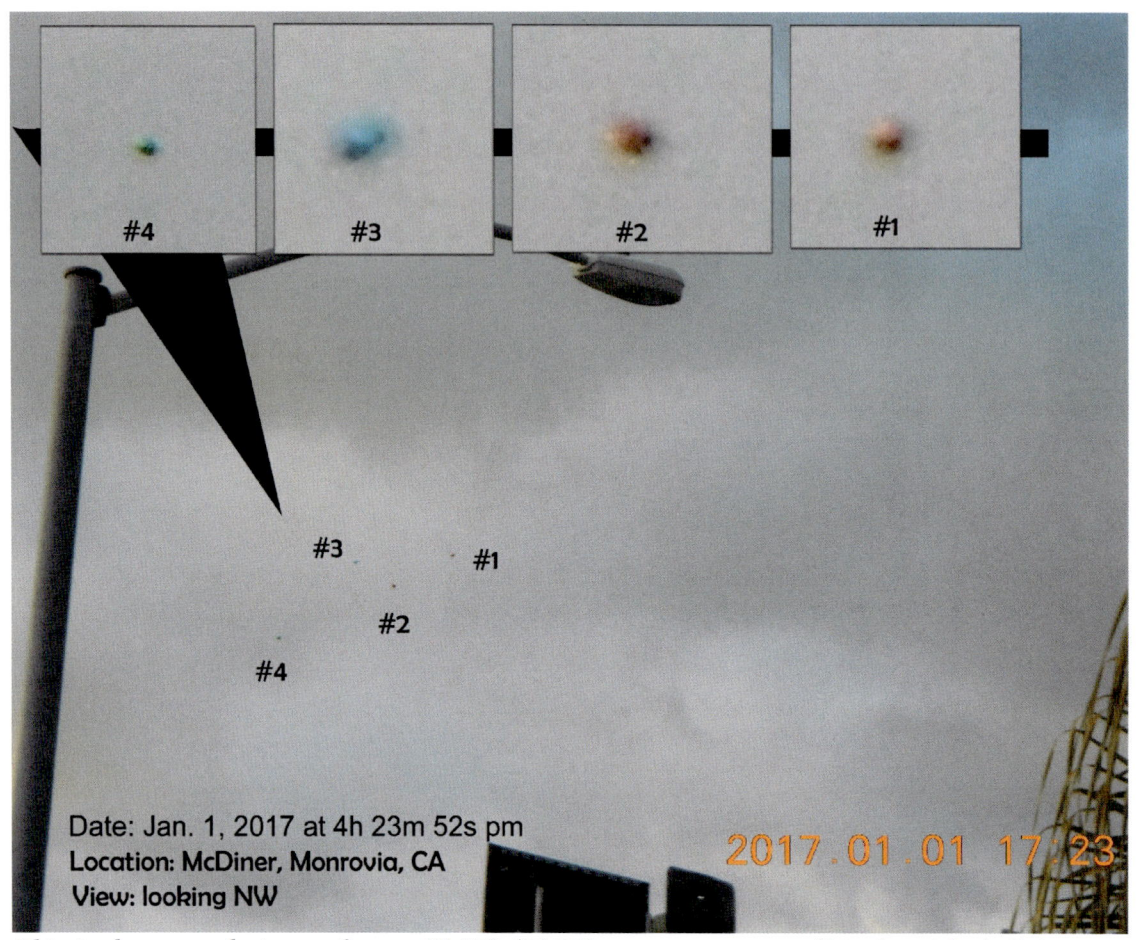

This is the second <u>picture</u> frame, #1679 (#1678, as you may recall is the video). It is shown here now labeled, and with enlargements of the craft included. This image is a definite improvement over the previous one. The orientation of each craft is now evident, and the splotches, which are just out-of-focus version of the panels, now are easily visible. Although there are color differences, it is now evident that these are all the same type of craft. (Montage version of image #1679 taken at 4:23:52pm with Coolpix S3100 camera; ©2017 Marian Rudnyk)

We now skip ahead to this, the forth (& last) UFO picture frame, #1681 (because we already saw #1680 earlier, with the beautiful detailed disk picture). Here #1681 is now labeled and with enlargements of the craft both as a set, and even more enlarged separately, also included. Once again hints of details are present. Also notice how these and previous frames all show each craft as having some sort of an "aura" around each one. (Montage version of image #1681 taken at 4:24:12pm with Coolpix S3100 camera; ©2017 Marian Rudnyk)

But Wait – There's More… A 5th UFO!

As it turns out there was a fifth object, which I didn't see at the time, but is clearly in the pictures. When showing the pictures to a friend, they noticed a fifth craft. This disk was in the distance, and approaching the other four. In the pictures it is visible on one side of the lamppost in one, and the other side of the lamp post in another. For me, at the time of the sighting, it was evidently hidden behind that lamppost, but fortunately, with the position I was holding the camera, the angle was just enough to catch it in the pictures. If you looked closely at the previous pictures, you may have noticed a tiny "#5" already marked in. Take a second look back, it's there. In any case, here are those two frames again, but with this fifth craft marked in and enlarged.

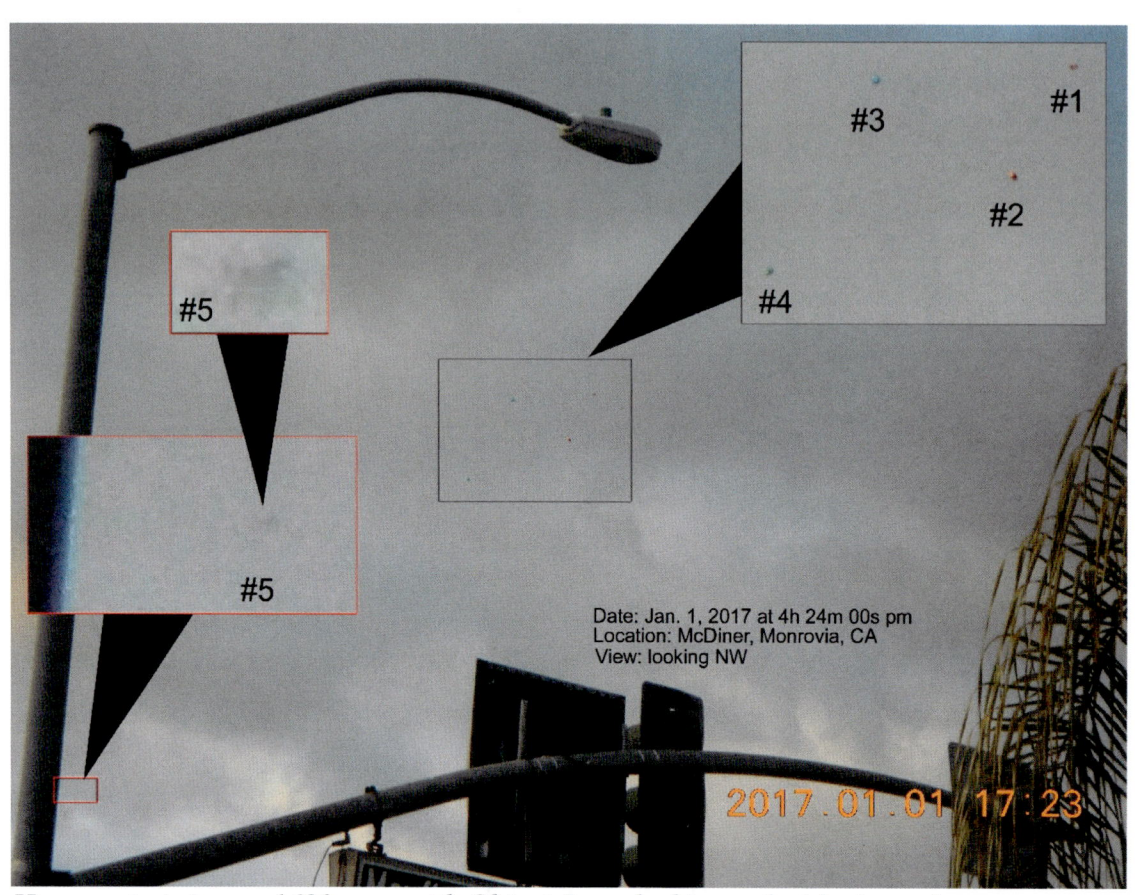

Here again is image 1680, now with Object-5 marked in, and with nice enlargements included. Object-5 is faint in this frame, but as you can see, it is definitely there. (Montage with object-5 version of image #1680 taken at 4:24:00pm with Coolpix S3100 camera; ©2017 Marian Rudnyk)

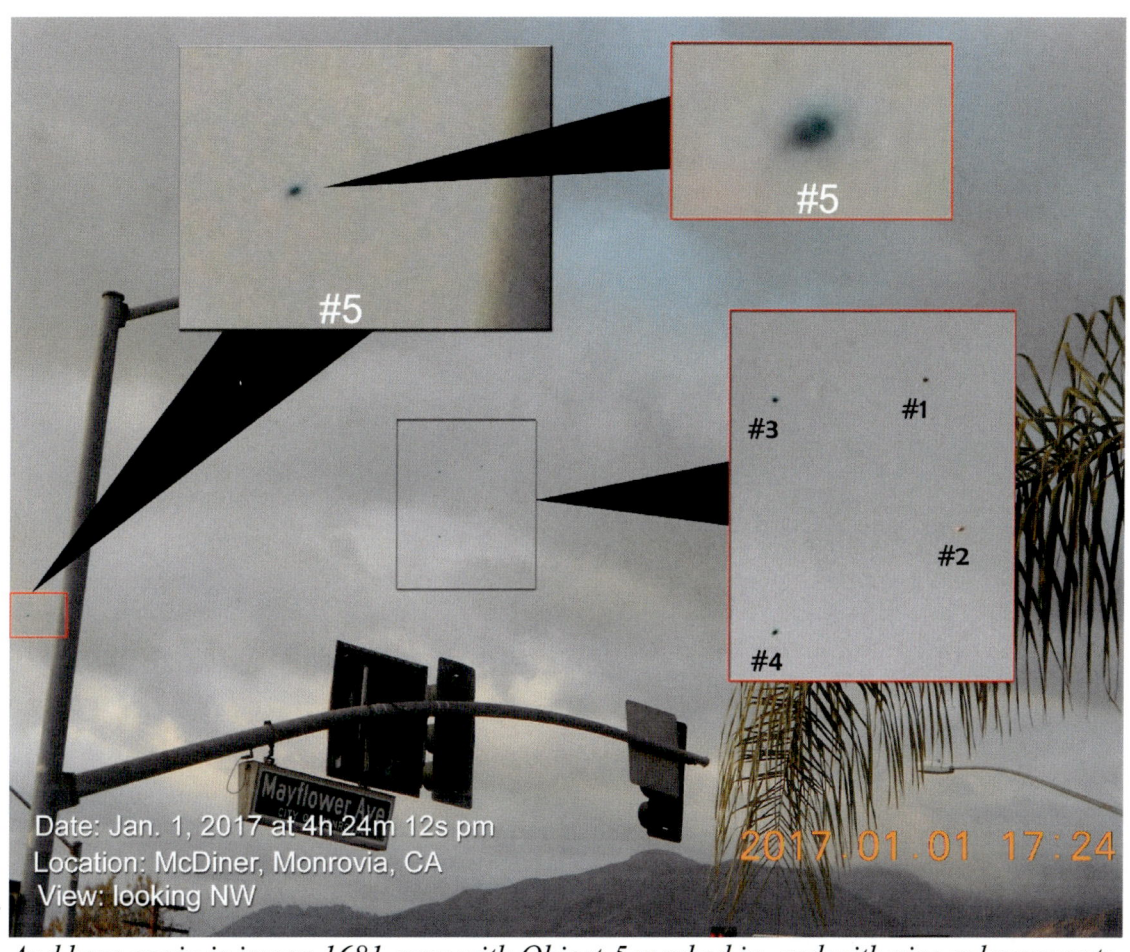

And here again is image 1681, now with Object-5 marked in, and with nice enlargements included. As you can see, Object-5 is easily visible in this picture. (Montage with object-5 version of image #1681 taken at 4:24:12pm with Coolpix S3100 camera; ©2017 Marian Rudnyk)

And here is a montage of close-ups to help you see and understand Object #5 better.

OBJECT #5

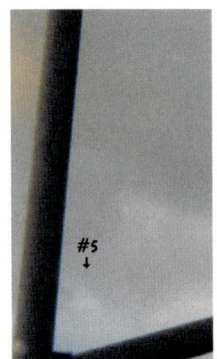

Frame #1680   Frame #1681
Note: frame #1681 resized so that both frames are matching in scale.

Both scaled frames combined along pole so as to show movement from A to B.

CLOSE-UP

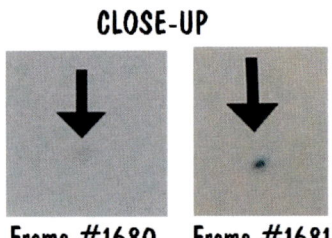

Frame #1680   Frame #1681

Object #5 was moving SE towards the other objects, and thus appearing bigger, and higher in the sky, the closer it got.

ENHANCED CLOSE-UP

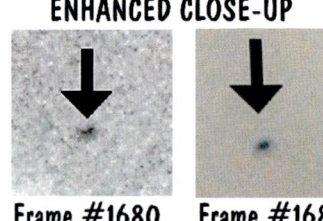

Frame #1680   Frame #1681

From this montage of close-ups we can see how Object #5 was moving in from the distance in the northwest in a southeasterly path taking it to Object 1-4. This is supported by both its position in the sky, and the fact that it is brighter and bigger in the second frame. At the rate of its motion, its safe to say that by the time I was taking my last photo, Object #5 was probably somewhere overhead. (Oh! Too bad. If I had only looked up directly overhead!) (Object-5 multi-frame montage; ©2017 Marian Rudnyk)

To get a sense of context and how this all fits together, here is a sketch I did of the area of the sighting at the time:

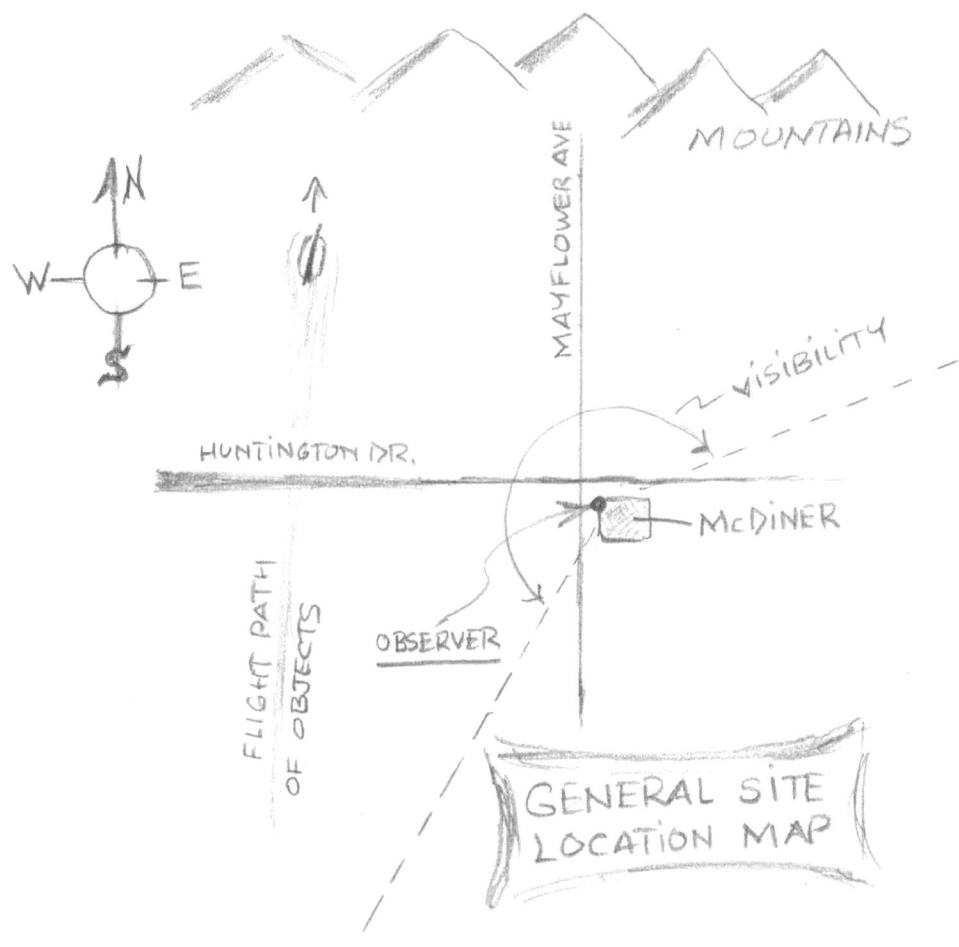

Area map quick-sketch (left) that I drew of the area of the sighting just after the event. (Area map quick sketch; ©2017 Marian Rudnyk)

As I had stated earlier, during each stage of this event, as details revealed themselves, I would update my definition/name for these "objects". At this point, one thing now became abundantly clear to me: these were not UFOs (Unidentified Flying Objects), and simply "craft" wouldn't do. These were, safe to say, more specifically "Unknown Flying Craft", or what we'll now call UFCs. (A little later in this book I will explain and discuss why I evolved these names, but for now please bare with me, the main point is basically to be more specific as I know more and more.)

What Does It All Mean?
Although it's now perfectly clear that I saw an interesting variation of the classic "flying saucer", or what I'm calling, for now, UFCs, let's take a deeper dive and learn some more from what we now have and understand.

First off, to get a better grasp of these craft, I think visuals are very important. Here then are all 5 craft, now with grouped enlargements.

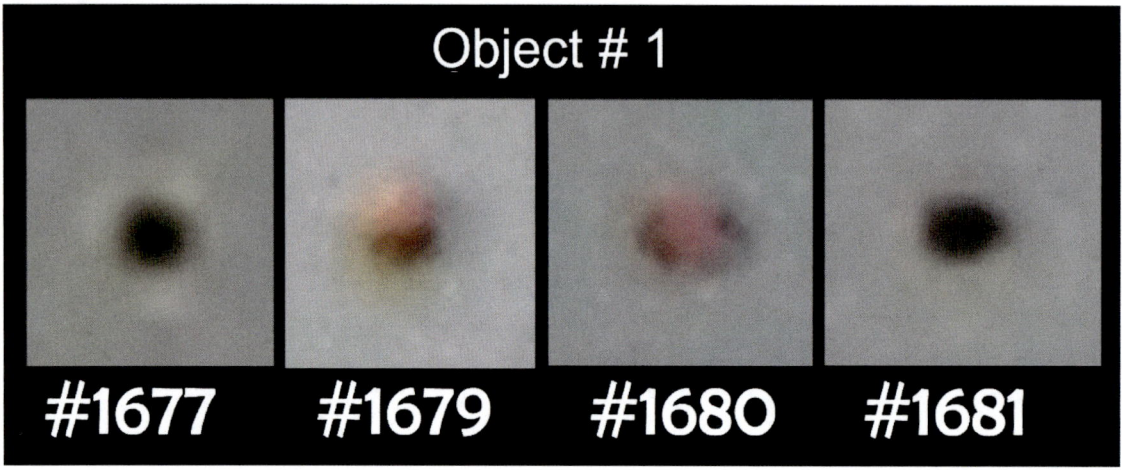

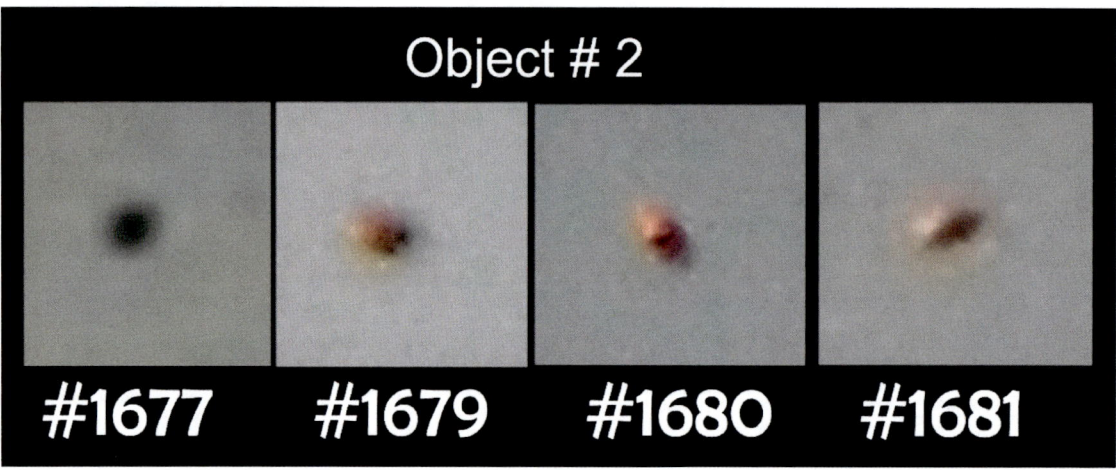

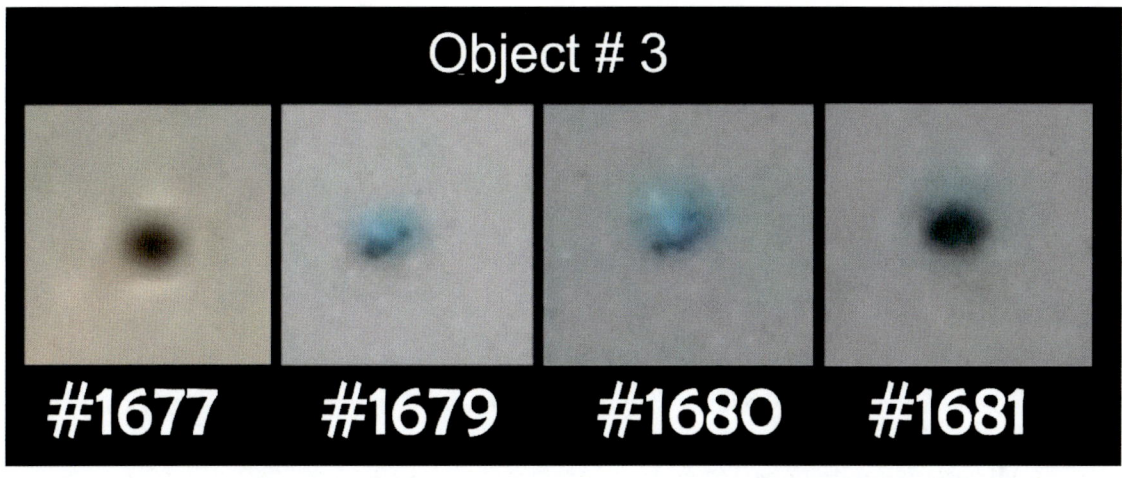

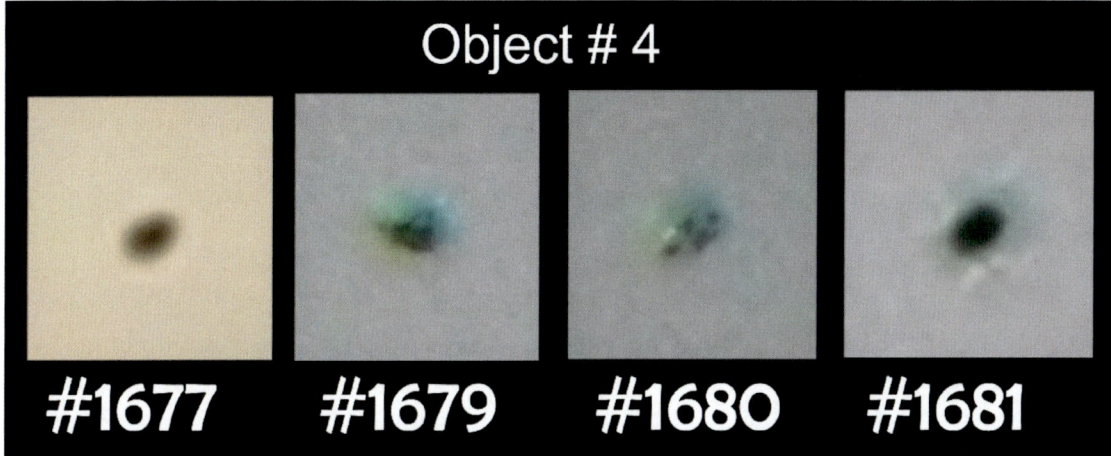

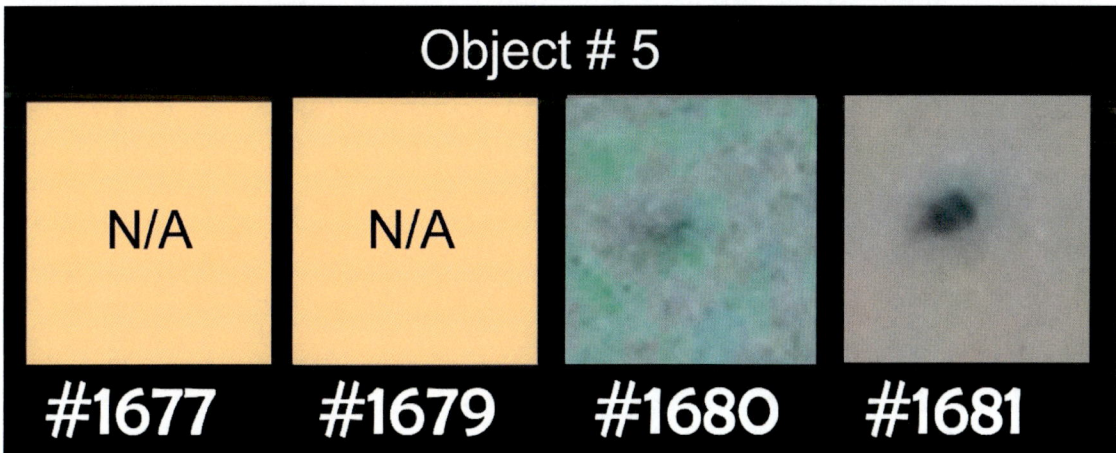

(Objects 1-5 image montages, all ©2017 Marian Rudnyk)

By grouping the UFOs in this way, several things immediately jump out at you:

1. All 5 have the same shape: are disks.
2. All 5 probably have triangular panels on their undersides. True, some of the images are distorted, but with Object-4 as a reference, it is easy to see that the splotches in the other pictures are most likely indicative of the same triangular panels, just a bit distorted – but nevertheless there.
3. They have different colors. Objects 1 & 2 are reddish, Object 3 is bluish, and Object 4 is greenish. Object 5 was too far away and indistinct to define a true color. We'll get into what these colors may mean in a minute.
4. None appear to have external flight surfaces (such as wings, propellers, etc.) of any kind.
5. All 5 show no evidence of engines or engine exhaust.
6. From the video, and my on-site observation, we know they are silent - meaning whatever propulsion method they use, it is noiseless.
7. Also from the video, and my on-site observation, we know that they appear to be immune to weather such as wind and rain. (Lightning, however, may be a different matter altogether.)
8. From the combination of the pictures, video, and what I saw, we know they are capable of great bursts of sustained speed. For example, Object-5 covered a great distance (miles) in a matter of seconds.
9. Lastly, for now, we can also safely infer that they are not infallible, as evidenced by the fact that Object-4 was struggling, and thus possibly damaged. Judging by the weather and conditions at the time, and the fact that these craft all dropped out of the heavy clouds that dominated the sky, it is highly likely that Object-4 was possibly damaged by a lightning strike. The only other possibility would be that the craft suffered an unknown malfunction of some kind. Whatever the case, these craft are not perfect, they are machines, and can be damaged.
10. We know for a fact, beyond the shadow of any doubt, that these are not some sort of digital artifacts caused by the camera. They are real: I saw what I photographed and video'd. But just to be clear, a digital artifact, such as a burned out CCD chip pixel, does not move, by definition. It appears in exactly the same place in every frame. These move frame to frame, perfectly consistently, as well as in the video. So-called "digital artifacts" also don't move across in videos. Therefore, the camera was obviously recording a real event – what I saw. Humans can sometimes doubt their own eyes (i.e. "…I can't really be seeing what I think I am seeing…?"). On the other hand, cameras don't "judge", they just record what they see. And most importantly there is the human factor: I actually stood there and saw these craft and actively filmed them. I had no doubts. I know exactly what I saw and had plenty of time, relatively speaking, to not only photograph AND video them, as well as simply observe. As a scientist and a highly trained observer, I know what I saw, and the camera simply recorded and confirmed what I was observing. These are all extremely important points to remember.
11. We know they were not any kind of "natural" creature. They were obviously not birds, bugs nor anything else, living or otherwise. These were machine-objects (craft), and not anything "living" in any sort of way.

12. Lastly, they were not any sort of natural or weather phenomenon. They were not Venus, Jupiter, the moon, dust devils, twisters, space dust, odd clouds, nor swamp gas. These were clearly defined craft that moved in a calculated and controlled intelligent fashion.

Taking all this into account, let's see if we can nail down what these objects, are not:

1. Airplanes
2. Balloons
3. Blimps, dirigibles
4. Drones
5. Helicopters
6. Rockets
7. Meteors
8. Digital "artifacts"
9. Natural phenomenon
10. Flying animals, etc.

Now that we know what they are not, what can we say they *are*? We know they exhibit the following characteristics:

1. They moved independently (no wires, strings, ropes, tethers, etc.) and against the wind.
2. No human element was present. In other words, there were no people in the area apparently somehow controlling them. No antenna of any kind appeared on the objects. And the objects flew using technology unknown to us. Thus, they were probably piloted.
3. They moved intelligently and under self-control. They did not exhibit any kind of random movements you see with balloons, and the like. I cannot stress this enough: they moved deliberately and with intelligent control and purpose, and were not affected by the wind.
4. They were not just "objects" (i.e. like a rock is an object), but rather they were machines – craft – that moved under self-control.
5. They were silent. There was no engine noise, nor any other kind of noise of any kind, put out by any of them. Yet somehow, they flew.

Without endlessly and needlessly belaboring all these and other points, it becomes abundantly clear that these were not simply objects but actual craft – vehicles, if you will - machines capable of flight using technology that is completely unknown to us. They are also not any kind of craft from this, nor any other country, since no one possesses this type of flying technology.

Which leaves us with one inescapable answer: these objects, actually highly advanced flying craft, were indeed some sort of non-terrestrial craft.

After doing some analysis and crunching some rough preliminary numbers, here's what I can say about these craft. (You can see the full details of how I arrived at this in "Appendix 3: Doing The Math".)

The best view we have of these craft is Craft-4. This is obviously because it was damaged, and was therefore the lowest. Thus, fortunately for me, the camera focused on it and captured this wonderful image (below). And reiterating, the Craft-4, like all the others, shows symmetry of features, thus supporting the notion that the splotches on the middle right side represent some sort of damage. Based on the weather at the time it is reasonable to assume that these are most likely some sort of burn marks from at least one lightning strike (or possibly more).

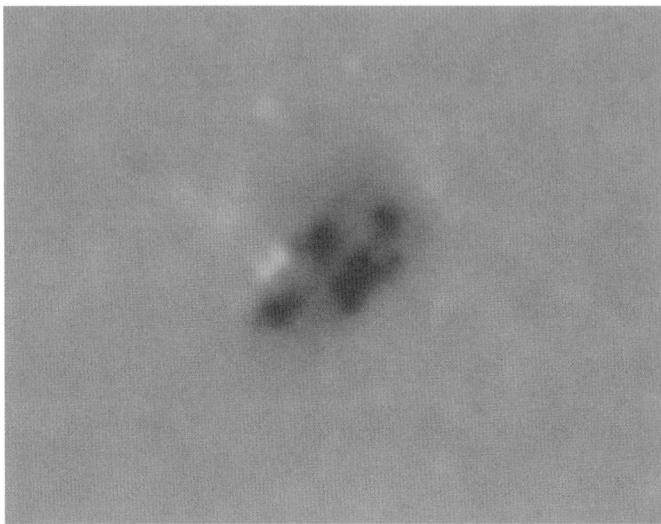

This is a close-up of Craft-4 from image #1680, it offers the clearest view that is representative of what all four craft look like (minus the damaged area in the middle right side). (Craft-4 close-up from image #1680 taken at 4:24:00pm with Coolpix S3100 camera; ©2017 Marian Rudnyk)

If you take the above image and geometrically re-project it and flatten it, we get the following approximate representative image.

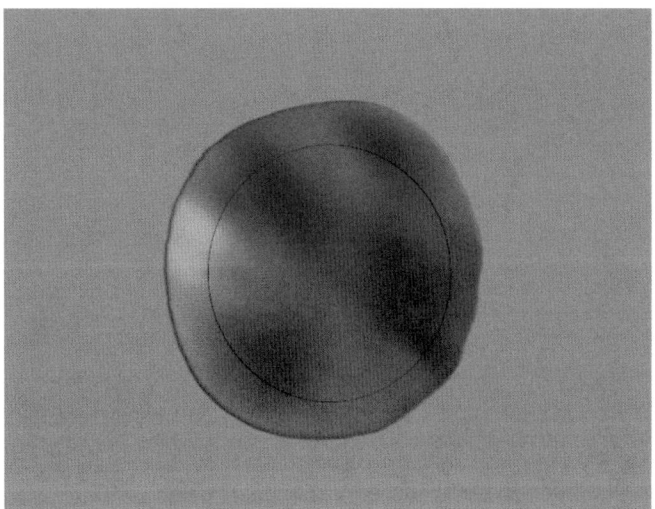

Here we see the image of Craft-4 removed from its cloudy background, geometrically flattened, and then placed on a neutral gray background. (Craft-4 close-up geometric re-projection from image #1680 taken at 4:24:00pm with Coolpix S3100 camera; ©2019 Marian Rudnyk)

To better understand the above image, I have done an artist concept quick-sketch, and added some graphic notes so as to help us visualize what everything means.

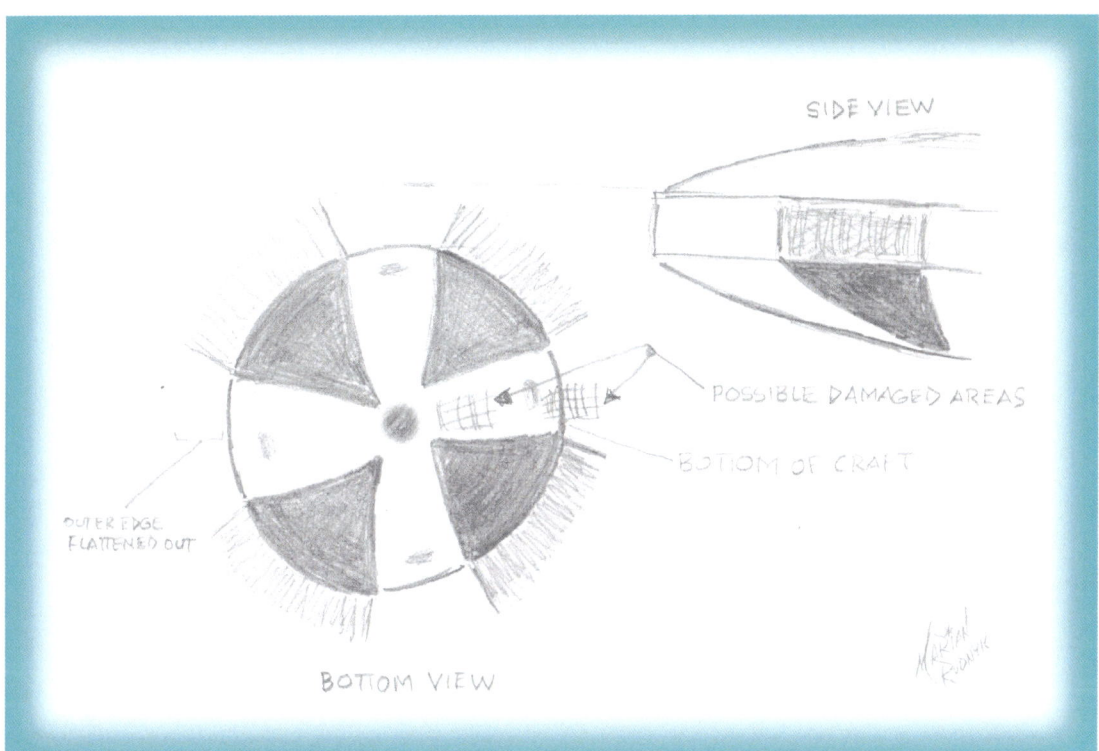

Sketch based on the flat re-projection of Craft-4, and my own recollections. (Craft-4 visualization quick-sketch; ©2019 Marian Rudnyk)

As can now be seen, the apparent symmetry of the craft's design, makes it very possible that the aberrant splotches on the right side (indicated as cross-hatched areas) could indicate damaged areas – possibly from a lightning strike. Craft-4's distressed flight behavior would tend to lend credence to this possibility.

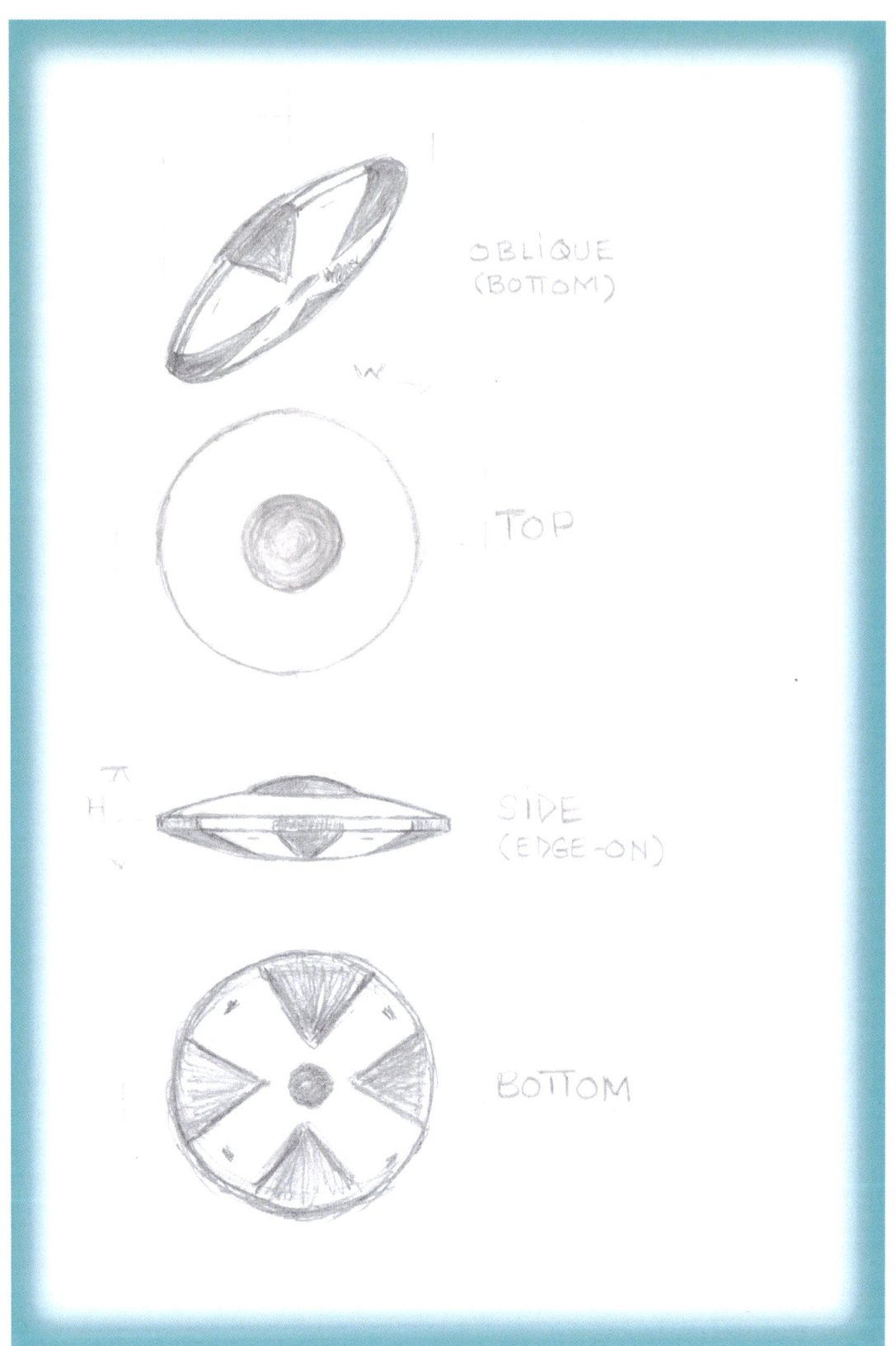

In this concept sketch I've diagrammatically laid out what we know about the design of these craft. Sketch based on photos and recollection. (UFO craft design sketch montage; ©2017 Marian Rudnyk)

Now that we have a pretty sold idea of what these craft look like it's time to put some numbers behind what we now know and gain a deeper understanding of these craft and the incident as a whole. The basic thing that I think obviously beg for answers is exactly how big is the craft. Yes, from my observations, the feeling I got was that each craft was about the size of a large car or so (if it was round). Just glancing at the pictures you get the sense that that is approximately right. So how tall (thick), specifically, is it? What is its exact diameter? How would the numbers stack up? Is there a way to tease out the numbers from the information we have so far?

The answer is obvious: geometry to the rescue! We need to put some numbers behind what we see in the pictures. Thus I decided it was time to map everything out, and then start measuring. In other words it was time to do some fieldwork.

I headed back out to the McDiner and the surrounding areas. My goal: scout everything out, take detailed measurements, and then crunch the numbers and see what I can come up with. First thing I started with was the creation of a map of the general area, to mark out what exactly was visible in my photos and video.

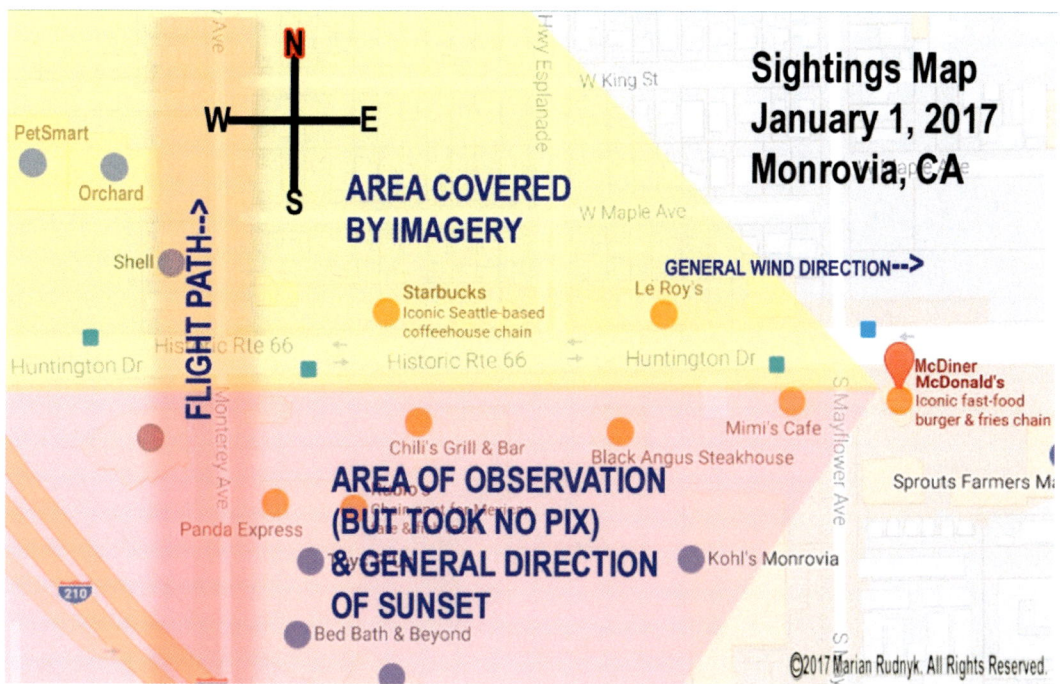

Graphic map indicates exactly what areas and landmarks were captured in my 4 main photos and the one video. Other pictures taken before and after the actual UFO encounter event itself do cover this area in all directions, including the sunset (see "Appendix 1: Photo Archive" to see the full set), but for our purposes all we need to worry about is what is visible in the yellow area. (Sighting Area Map ©2017 Marian Rudnyk)

Now it was time to really get into the nitty-gritty. I compared my photos to what I could see, identified landmarks, and then went to work measuring. The result was the following highly detailed map. The goal was to assign numbers to all the landmarks seen in the photos and video. This would then allow me to calculate, with any luck, the approximate size of the craft and any other interesting information I could tease out of the numbers.

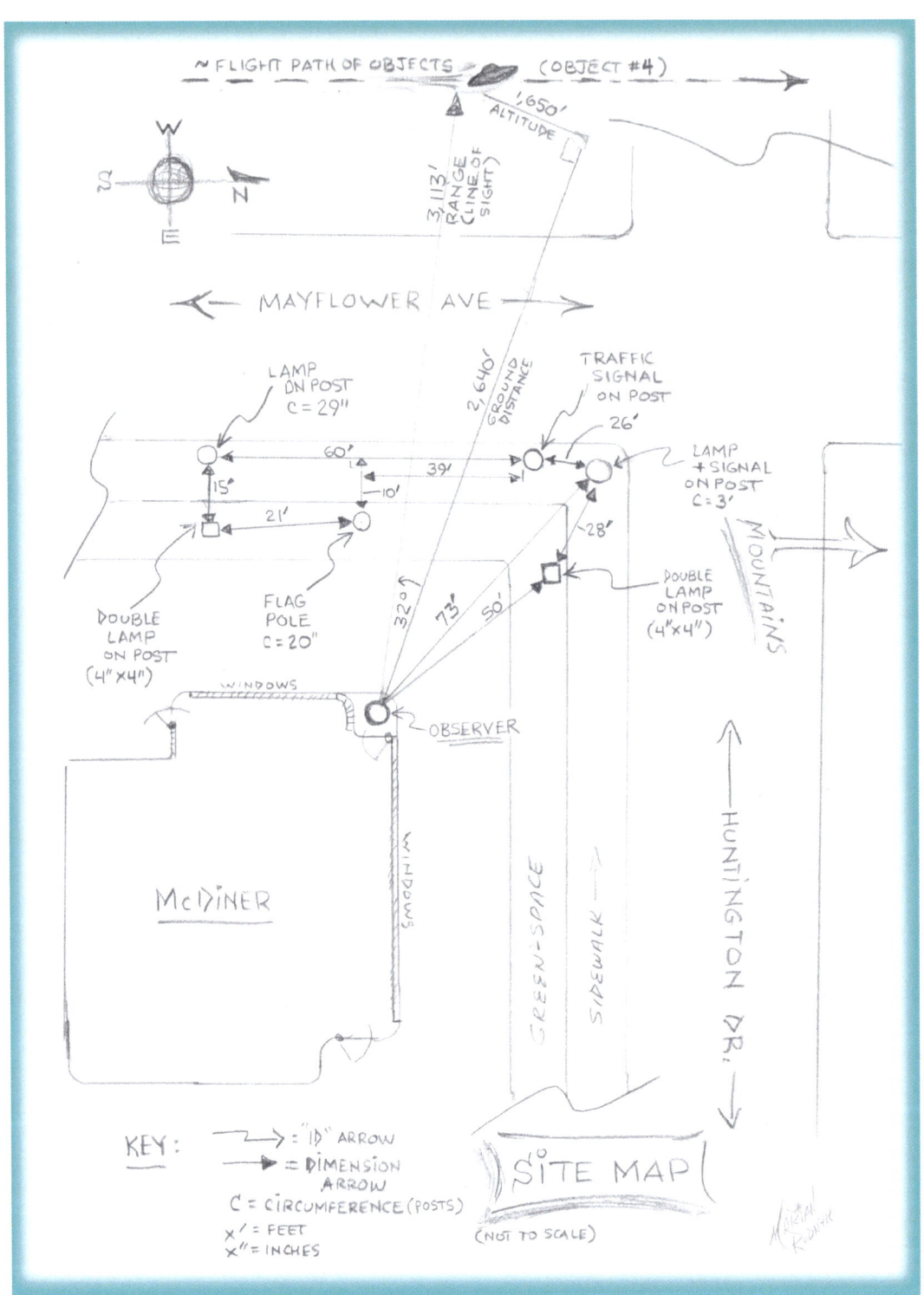

Sketch map of the area, now with highly detailed measurements that I took. (Detailed sighting area map; ©2017 Marian Rudnyk)

I now decided to take all I know and put it to the test. What could I figure out? Don't worry, for those who are faint of heart when it comes to math, I will spare you the gory details of my math. If you ever doubted the usefulness of taking a geometry class, here's proof of a practical application to all those angles and theorems your teachers tried to teach you. For those daring individuals who like getting into the weeds, you can see how I crunched the numbers in "**Appendix-3: Doing The Math**". For the rest of us, let's continue on with the intriguing results…

So here then, excerpted out of that appendix, are the results of what I learned about these flying craft:

DESIGN: Careful review of imagery reveals that all 5 objects are all artificially constructed craft of matching design. As discussed before, these are not natural phenomenon. Phenomenon would describe something of strange natural origin, while these are obviously artificially designed and constructed craft, or vehicles. They do not match any known configuration of contemporary craft anywhere in the world.

SHAPE: All 5 objects have bilateral symmetry: they are circular disks, and have the same generally curved, convex, saucer shape both top and bottom – much like, for example, the shape of a magnifying glass lens.

SIZE: Each disk measures about 18.4 or so feet in diameter and about 6 feet in height (base to top of dome). The ratio of diameter to height is about ~3:1. One can generalize that they are about the size of a large car (if the car was round).

COLOR: They may come in at least 4 different colors: white, blue, gold and red. No color is discernable for Object # 5 due to image quality. The panels on all of them appear to be greenish in color. The colors may be inherent, or they may be a product of reflection of light at certain angles of some sort of exotic or coated metallic surface. The colors may also be the result of ionization of gases. In other words, if the surface of the craft, either the saucer, or the panels, or both, is somehow charged, the surrounding terrestrial gasses may be reacting – much the same way that auroras are created in earth's atmosphere (when charged particles in the solar wind interact with the gases in our atmosphere). This ionization theory is especially supported by the fact that the colors of the crafts seem to have a halo-like quality to them in all the images. Therefore, as the electrical/magnetic field generated by the craft, or the charge of its surfaces changes, so does which gas responds to this process. This halo effect may also be at least part of the reason some of the images of some craft are distorted. All this may actually may provide insight as to the workings of these crafts' propulsion systems. Finally, with regards to actual color, this means the craft themselves may actually be whitish (possibly metallic), and have greenish (or some other dark color) panels, as evidenced in Object 1, for which exists the clearest imagery.

PANELS: Each object has 4 triangular panels on their undersides. The triangle-base-edge of each panel partially extends around the perimeter edge of the disk. The panels overlay the smooth surface of the main shell body of the disk, and are possibly slightly raised over its surface. This orientation is very specific and most likely also a clue to the method of the crafts' propulsion.

BOTTOM DETAILS: Though difficult to see in each frame, each craft probably has a dark centrally located circular area on its underside. Although the bottom is curved, like the top, this area (the circular part) appears to be relatively flat. There also appears to be a small round, or rounded square area, what I call a "ports" (just for the purpose of convenient description) located midway between each panel, close to the perimeter. The existence of these ports is conjectural based only on the pictures, at the limit of their resolution.

TOP DOME: Each object has a low central dome on top (as hinted by the photos). The dome has a defined double-border edge where it sits on the main saucer body.

FLIGHT: Because they are flying in a variety of orientations, yet all generally maintaining formation and speed, wind apparently has no affect on them – at least not the gusty type observed during the sighting. The performance of the craft in higher, or severe winds, is obviously beyond the scope of the observations of this sighting, and therefore indeterminate – but should be noted for future observations.

Additionally, it can be concluded that they are capable of complex formation flying, flying at controlled slow speeds, possibly hovering, and also great bursts of speed when needed. They thus demonstrate a great versatility of complex speed and maneuverability. During the sighting Objects #1-4 cruised at an average speed of 25 mph. Object #5 showed a massive burst of speed between frames #1677 and #1679, and then cruised too.

PILOT: It can be inferred from their size, that these craft are probably manned by a single pilot, maybe two. They may also be remotely piloted UAVs (Unmanned Aerial Vehicles) of some sort. The presence of a dome on the top half of the saucer tends to support the notion, however, that these are not robotic craft, but are indeed piloted craft.

PROPULSION: Based on their design and flight characteristics, these craft show no currently recognizable form of known propulsion. The obviously regularly accepted general principles of aerodynamics seem to not apply – but that is probably only because the technology is not outwardly apparent at the present time. The distinct shape and the layout of the panels are probably clues to the technology used.

SOUND: The craft move silently. They produce no audible sound, at least not within the range of the human ear. Whatever the propulsion used, it is "apparently" soundless.

ORIGIN: The craft are either 1. U.S. (ours/experimental), 2. Foreign (terrestrial, undetermined non-U.S.), or 3. Outsider (extraterrestrial, or of unknown origin). To be honest, at the time of the sighting, I had no idea how to answer this question. I felt all three of these possibilities were equally viable. But that would quickly change. Within only a matter of weeks I would know exactly the answer. The heavy-handed military response would be the biggest clue. Yes, these were definitely extraterrestrial craft – not of this earth.

In Summary

So what do we know so far? These craft are each about the size of a large car. They are disk shaped, and about 18.4 feet in diameter, and about 6 feet high at their middle. They are probably piloted. They can move slowly, hover, and are also capable of explosive bursts of high speed. They each have 4 triangular panels on their bellies, and these may be related to their propulsion system. The pictures hint at the top of each disk having a central dome, presumably where the pilot(s) sits. These craft are probably some sort of light metallic color. The colors that they exhibit are probably due to some sort of ionizing glow. The different colors of this "halo of color" probably hints at how they are able to fly and what the propulsion system is doing at any given moment. (Although I've commented throughout my account so far, about how these craft moved, if you're a fan of "detailed analysis" then feel free to check out the last section of Appendix-2, where I do a deep dive on this topic, and even have a nice graphic that plots out how the craft moved and how I arrived at my conclusions.)

So we can now summarize all this by saying:
The objects were disk shaped, about 18.4 ft. in diameter, and about 6 ft. high in the middle. Had unique triangular panels on their undersides. They can maintain slow speeds (25 mph or less), but have the ability of high-speed bursts, or more, whenever they choose. They flew about 1/2 a mile away from me. They were at an average altitude of about 1/3 of a mile. They flew in a controlled deliberate manner and were thus not just "objects" but intelligently piloted craft. Because they were probably piloted, we can go so far as to call them vehicles. There were a total of four craft that I saw (and a fifth revealed in the pictures). Of these, one was apparently damaged (most likely by lightning), and may have crashed in the foothills/mountains above Monrovia, California.

This is an amazing amount of information. Still, the most important takeaway here, so far, is that they are definitely "not of this earth", and are alien in origin. Their relatively small size hints that they may actually be some sort of scout/recon craft. Their technology is obviously far beyond any possessed by anybody here on earth.

Unfortunately, this is not the full story. The 1960's TV show about a hidden UFO invasion called "The Invaders", starts with the words "How does a nightmare begin?" For me, this sighting was just the beginning of *my* nightmare.

One More For The Road

But I had a secret – something in reserve that I shared only with my brother. On January 20th, barely 3 weeks after my initial sighting, I had had a second sighting – and it was from the McDiner again! And I had pictures to back it up. Not as nice as from my first sighting, but still it was something. Little did I know that this would all be a harbinger of the many crazy things to come!

On that day I saw a single bright disk shaped object flying along the foothills and mountains. Although it was heavily overcast, the object was very bright and moved at incredible speeds and in a very deliberate manner. It quickly disappeared into the clouds, but eventually reappeared again and flew directly south, very quickly, over the city (as can be seen in the pictures below). Then in a split second, in a massive burst of speed it shot up into the sky and disappeared.

This picture was taken on January 20, 2017, at 1:55:44 pm, at the McDiner in Monrovia. The view is to the NW. The conditions and time, are eerily identical to my first sighting on January 1st. The bright object did not fly in any way normal for conventional aircraft. Its flight behavior seemed to indicate that it was especially "interested" in the foothills and canyons just above my house. (Montage version of image #DSCN0002 taken January 20, 2017, at 3:55:44 pm with Coolpix S3100 camera; ©2017 Marian Rudnyk)

This picture was taken on January 20, 2017, at 1:57:12 pm, at the McDiner in Monrovia. Note that both pictures show a time that is one hour off (my camera had still not been reset for Daylight Savings Time). The view here is more directly to the N/NW. The object though appearing small here, was extremely bright, as evidenced by how it almost "burned in" the pixels on the camera. (Montage version of image #1788 taken January 20, 2017, at 1:57:12 pm with Coolpix S3100 camera; ©2017 Marian Rudnyk)

With these pictures not being as nice as my initial sighting, and the fact that I felt strongly that I already had enough problems dealing with one sighting, let alone two, my brother and I decided to table this one at that time. It is being revealed here for the first time ever for the sake of completeness. Was this object somehow connected with the craft on January 1st? I didn't know. But Ultimately, it proved to be part of a bigger picture – something I'll get into a bit later. And that now makes it important as well.

Soon the military would get involved. Bad things would happen. And more sightings would happen. Lots more! Things would escalate to dangerous levels. Yes, my nightmare was indeed *just beginning…*

3:

Aftermath: The Mystery Unfolds

Timing Is Everything - Almost...
What I like to affectionately call the 'main event', the actual sighting itself, happened back on January 1st, 2017. It begs the obvious question: why now? Why the delay? Why talk - now?

To be blunt, as I will detail in the coming pages, I was literally told not to 'talk' – sort of (I'll explain what I mean by "sort of" in a minute). However, as time progressed, my situation – and the related government harassment - became worse and worse. I soon realized that it didn't matter whether I talked or not. Those within the government with the authority and power to harass were not going to let up – regardless of what I did. By summer of 2017 things had gotten so bad that I decided to "talk", which in my case was to begin crafting a strategy to disclose what I knew, and that included this book you are now reading.

Then something changed. Almost a year later, "disclosure" happened on December 16th, 2017. It was like an off-switch hit. Everything changed. Suddenly my 'problems' all simply went away. A contact I had, reached out to me with this simple cryptic phrase: *"write your story –now- while you can"*. The inference was that things might easily go back to the way they were, but now –right now- was an opportunity... seize it while it lasts. And here we are...

Now, the so-called 'disclosure' event I mentioned previously, to those unfamiliar with what happened on December 16th, was the release of a previously classified DOD (Department Of Defense) UFO video (known as "Gimbal") by an organization known as the To The Stars Academy, or TTSA, and a front page article in the New York Times, no less, of the existence of a secret government UFO project AAWSAP ("Advanced Aerospace Weapons Systems Application Program", possibly also referred to as the AAWSA Program). This program is now commonly known as AATIP, or "Advanced Aerospace Threat Identification Program". This release was quickly followed by a second video (known as "Nimitz FLIR-1"), and finally, the next year, on March 9, 2018, of a third video (known as "Go Fast"), as well as recently declassified related documents. (As of this writing even more documents are expected to be declassified within the coming months and year.)

Now while all this may seem like a lot to digest to the uninitiated, don't worry, I will fill in the blanks as we continue – as best I can. For now, the takeaway is that it was a matter of timing, and that timing was, for the most part, out of my hands. Yes, I could have released this book last year... but that would have been ill advised – for a number of reasons I will touch on later.

Enter U.S. Air Force SPACE COMMAND
If you think this story ends with my sighting, you'd be wrong. Very-very wrong. I wish it were the end. As it turns out, the sighting was just the beginning.

What follows is like something from a dime store spy novel. I would be plunged into the world of Dark Ops, national security concerns, and outright shadow government harassment. Unfortunately, I remain mired in that world to this day. We'll touch on all that in the next chapters. Right now we're going to look into the stunning events that followed directly after my sighting and the shocking revelations, and inescapable conclusions, that would cause me to have an absolute paradigm shift in my thinking as it pertains to purported otherworldly activities in the skies over our planet.

My sighting happened on January 1st, 2017, and just days later on January 5, 2017 I got an email from my good friend Sarah inviting me to dinner. Three days later, on January 8, 2017, I had that fateful dinner with Sarah. Why fateful? Because I took with me a print of the picture of Craft-4 – a revealing close-up version that exposed the craft to be a saucer with panels and other details. I told Sarah about my sighting and explained how seeing these four craft was causing me to now have a paradigm shift in my worldview. She was curious, listened intently, and seemed sympathetic and understanding. She knew me very well, and had a deep appreciation for my expert background, so she knew I was shooting straight with her – more so than I realized.

Afterwards, I got a call from her saying that she needed some help. So a week later, on Sunday January 15, 2017, I came over to Sarah's house. She was eager to see the pictures and video from my sighting, so I showed her everything and went over the whole event in expanded detail. She was quite impressed. Afterwards, I proceeded to help her with some tech stuff and then her resume. While I worked she was on her smartphone on Facetime with her sibling.

As I finished she said she needed to briefly visit a nearby neighbor and said she would be right back. As she excused herself and headed off she placed her iPhone on the table, still connected to Facetime, and said I should talk to her sibling while I was gone. I shrugged and agreed, and she headed off.

As we introduced ourselves her sibling said I needed to meet their spouse, and put that person on – a command level officer with the United States Air Force's Space Command. We'll call that person "Mary". I will never forget that exchange. It went something like this…

"Hi Marian, nice to meet you. My name is Mary," she said.
"Hi. I recall hearing about you," I replied, referencing the fact that I had heard that Sarah's sibling had a spouse.
"So I heard you saw something. You have your laptop with you, right?" she asked.
"Yes, I do," I answered. "It's right here and still turned on. Why do you ask?"
"Why don't you pick up Sarah's iPhone, turn it around and show me what ya got?"
"Ok," I replied, thinking this was all-in-the-family, so no big deal.
I placed the iPhone so it could see my laptop screen and did a show and tell of my sighting. They were riveted. The reaction of Sarah's sibling and spouse was palpable to a level that caught even me by surprise.
The first reaction was from Mary. As I zoomed in and showed a close up of Craft-4, which offers the clearest view of the details of the disks, Mary commented exactly, "Yup, that sucker is real alright. They're all real. That's pretty amazing."
Sarah's sibling pushed Mary out of the way and started excitedly blurting out, "You need to go the press! You gotta call CBS! You gotta call CNN and FOX! You're an astronomer – they'll listen to you!"
"No. He's not going to do any of those things," said Mary stoically as she pushed Sarah's sibling out of the way. "You need to know that I work for Space Command," stated Mary. She then explained a little about her work and the need for secrecy. I said nothing, and just listened.
Mary's spouse then interjected again, excitedly telling me to ignore Mary and talk to the media – talk to anyone that will listen – and that because I'm an astronomer people will listen.
"That's right. You're right," interrupted Mary. "And that is exactly why Marian is going to listen to me and stay silent."
I watched them arguing, almost comically and with good humor, and I said nothing. I just smiled as best as I could – too stunned for words. Finally, Mary stopped the "comedy" and sternly said, "Marian, this is what you are gonna do: you're not going to talk to anyone, not the media, not anyone. And you're going to send me copies of everything you've got. The pictures. The video. Anything else you can think of. I want copies of everything."
"Ok," I answered, "I'll send you copies of what I have," carefully choosing my words so that the only thing I technically agreed to was sending Mary my data, and nothing else. Fortunately, because of Mary's spouse's interruptions, Mary never noticed that I hadn't agreed to stay silent, although, in hindsight, I'm sure that that was what he thought I agreed to. To be absolutely clear: did he tell me to stay silent and not talk? Yes. Did I say I agreed to not talk? Absolutely not. Did I ever sign anything, an NDA (non-disclosure agreement), or any other document? No. Did I ever promise in any email or any other form of correspondence that I would not reveal the things I am revealing in this book and in the many "discussions" I've been having? No. Did Mary assume that I would not speak? I think yes. Should Mary have sought something more exact and formal? Perhaps,

but then, that was not then, nor is it now, my responsibility. My responsibility is as a patriot who cares deeply about his country and the importance of the effect of the truth on our great nation. So-called "dark forces" have conspired to keep the truth from the public. This is not the American way.

In the meantime I consulted with my brother. Suddenly the kid brother I joked as being the family "UFO-freak" was the only credible resource I could trust. My concern was that if Space Command and the USAF (U.S. Air Force) confirmed what I knew was true once the reviewed the copies of my pictures and video, I could be very specifically "shut down" and told not to say anything – no holds barred. Even worse: my sighting could become classified. That would be a disaster.

Whatever the case I needed some answers, and fast, before the doors slammed shut on me. I had a multitude of questions, but number one among them was: were these craft ours or really from beyond earth? My reasoning at the time (remember, this was all new to me then), was naively that perhaps these were some sort of military experimental craft.

If that wasn't enough, my mind was a swirl of thoughts: everything from my past suddenly came flooding back to me. No longer was it all in the proverbial rear view mirror. It was all now *front-and-center*. It was time to seriously re-assess my "reality".

As a kid I had self-conceived the idea of ancient astronauts, felt that I had it confirmed by Von Daniken, and then simply moved on. These were "ancient astronauts", after all, so except for the historical relevance, I so no pressing connection to the present – nor any sort of perceived threat of any kind. The whole UFOs and aliens are in the here-and-now was more science fiction to me. And life simply had gone forward in the warm fuzzy earthly cocoon of a world that had yet to set foot, so-to-speak, in the "galaxy proper".

Now with my sighting barely a few weeks old, I was being forced to not only accept a paradigm shift in my world-view, and universe-view, but to also adjust my life accordingly. And this time it was the scientist part of me that was saying "wake up, and take action while you can". Piled on top of that was the nagging tugging of my heart reminding me, "if this is the truth, are you willing to just step aside and see the proof buried?" That answer was simple: a resounding NO. But what exactly was the truth? I knew what I saw. I also had pictures and a video. What I was missing, for lack of a better word, was "context". I needed answers.

My brother told me there were only two people who could really help me: Stanton Friedman and Bob Wood. He explained that they were beyond a doubt the most pre-eminent living UFO researchers in the world. Both, he explained, had serious credentials in the aerospace, scientific, and military communities, and were highly respected - in spite of the research they did in what he called the "UFO phenomenon". And in the UFO community, unlike the many nut-cases (and he admitted that there were plenty), he emphasized that these two had legendary status even in the UFO community, and among what he called other "ufologists" – a real rarity.

I then told my brother about my previous "encounter" with Stanton Friedman, and his words to me were basically that I needed to "suck it up" and reach out to him – when the time came. He explained to me the impeccable reputation of each man. He stressed that both would appreciate my background and be able to help me.

Decision Time – What Would I Do?
After much discussion, however, he came to a shocking conclusion: my brother actually strongly cautioned me against contacting either one. He was concerned with how the government might react, promise or not, if they knew I leaked out the news of my sighting, even in a confidential private setting, with one of these two legends. Ultimately, his advice boiled down to something like: see what the government says first, then decide what you want to do. My concern, without telling him at the time, was that the government would make the decision for me, and I was almost positive I knew what that decision would be, and that –if it came to pass- it would be something I would not like. My sighting would be classified… or worse.

I agonized over my decision. With all of that in mind, against the advice of my brother, on January 18, 2017, barely two-and-a-half weeks since my sighting, at 3:59am I secretly sent an email to Stanton Friedman with the subject "*Astronomer Needs Your Help*" (see "Appendix-4: Support Materials" to read the full email). I explained I had a sighting and that I needed help. To my utter surprise, he replied almost immediately, at 6:06am. Ultimately, we agreed to speak – that same afternoon at 2pm! I was utterly sleep deprived, but extremely excited to talk to him.

When I phoned him I found him to be a very friendly and affable fellow! We immediately hit it off. I set the tone right away by sparing no time in explaining to him that we had actually met once before during my time at NASA, at the Jet Propulsion Laboratory in Pasadena. He listened intently to my story of how we crossed paths and "swords", and how I didn't know who he was back then, and of my impressions of him. I explained what my brother had told me about him and that that was why I was reaching out to him. I had no idea what reaction my admissions would get from him, but I felt strongly that it was important to start with an honest and clean slate and be as upfront as I could. To my surprise he simply chuckled and said he recalls being at JPL back then, but didn't remember the encounter specifically because evidently it had gone so smoothly and without incident. He then chuckled and admitted that he often showed a gruff exterior to people, and knew he rubbed many the wrong way, and that it was no big deal. With that out of the way, we both shared "war-stories" about our times doing various space related work, shared personal anecdotes, and found we had many many things in common including ethnic roots. As it turned out we spoke for several hours! I sent him copies of all my materials, and we reviewed everything. Additionally, he vowed not to reveal anything without my specific approval – and much to his credit - it is a vow he has honored to this day. During our many conversations and email exchanges he implored me to reach out to Bob Wood, because of his military expertise, and provided me with Bob Wood's direct contact information. He stressed that the strong military interest Space Command showed in my sighting warranted a military analysis of what happened. He explained that it could help shed light on events and help me with the answers I sought – along with how to proceed in the future. He also was very clear that Bob Wood, like himself, was someone I could completely trust. My interactions with Stanton Friedman, and his staff, helped me take a fresh critical analytical look at my sighting and eliminate any other explanations for what I witnessed. If I had any even tiny lingering doubts about my sighting, the brutal assessment we did removed every last sliver of doubt. As it turned out, it wasn't so much the validity of my sighting that was in question, but the implications of what I witnessed. Key among concerns turned out to be the fact that Craft-4 had been struggling during my sighting. Now, more and more, I started to see why the military was so interested. Talking to Bob Wood would now prove critical in understanding even more.

But Things Were Now A Whirlwind!
On January 20, 2017, only three days after my momentous phone call with Stanton Friedman and the numerous interactions afterwards, and barely five days since my Facetime meeting with Space Command at Sarah's house, I got an email telling me Space Command is still waiting for my materials. Bottom line: I was worried and stalling. I knew that handing over all my materials to Space Command might cause a reaction once they concretely validated my sighting. I felt I had to act while I possibly still could.

Things were coming to a boil and I still felt I needed one more thing: insider perspective from the DOD world. While I continued my interactions with Stanton Friedman, I decided to simultaneously reach out to a friend of mine, Jane. She had once worked at NASA, and was now involved deeply in military space work. And, for now, for the purposes of this book, this is all I can tell you about Jane.

When I called Jane up, the conversation was short and crisp. Several days later she came over. She asked to see my materials, and so I did a nice little show-and-tell. When I was done she sat there stone-faced.
"Are they real?" I asked.
No answer.
"Are they ours?" I then nudged.
Nothing but a stone face.
"Are they, uh, NON-TERRAN?" I finally asked, using the term for the first time in this type of context, in I don't know how long.
She gave me a blank look.
This was ridiculous. I wanted answers, so I decided to shake things up. Next I finally added, "Hmm. You know - they were flying low enough that if they ever returned perhaps I should just try to take a shot at them. Then I'd REALLY have proof."
That finally got a reaction, "Definitely not," she simply replied.
Now we were talking, barely, but talking. My impression was that even Jane wasn't sure of their origin, but might be able to find out – and not in a way that would do me any good. I didn't know what to think.

The coming days were crazy. Shortly afterwards, Jane appeared on my doorstep a few short weeks later (after my first informing her of my sighting), with four books she wanted me to read asap - and possibly before I did or decided anything. She wanted my opinion on each one. One dealt with the history of UFO sightings. The next was about the consequences of disclosures. The third was, lets' say, speculative about certain types of physics, and the fourth was the same but even far more speculative. As Jane put it: *I should consider book one as relatively true, the second as things I should be considering, the third as inspirational science and engineering and what do I think of it, and the last one as total bunk, but is there anything there that I thought was worthwhile.* For now I'm going to hold off on identifying the books, except to say that this whole exercise had helped me get a foothold on everything I was dealing with – big time. In her own way, Jane was helping me. True, I still felt lost, but now it was a grounded sort of lost. If nothing else, I now had a grasp of the history of this stuff, and the key for me was to fit it all into my so-called puzzle and tease out some answers.

However, on the other hand, Jane was obviously not telling me everything, and I knew it. And I could honestly appreciate that. She was in a tough spot: she had a life and a family and couldn't risk breaking classified rules to "help" a friend, no matter how legitimate it might be. And she never did. But she could help me realize that I should stay quiet. And she tried.

What Jane didn't know was that I had already spoken to Stanton Friedman, and that had wetted my appetite even more. I was now also itching to speak to Bob Wood too, as Stanton Friedman had suggested. But Jane wasn't the only one in the dark. My brother didn't know either. I decided that I had to figure this out on my own. All these interaction represented pieces to a much bigger puzzle to me – a puzzle that I was determined to solve. But intersects are funny things, just as you think you're in control, a new one pops up, and you get your answers – but the cost is a headlong plunge into a whole new direction and world. I was about to take that plunge – a plunge into the dark world of UFOs and much worse. And the answers I got raised more questions – each with profound implications.

Things Escalate & The Military Makes Its Move!
On February 1, 2017 Sarah emails again. Space Command is waiting! They are obviously chomping at the bit. I tell her that I lost their email, so she gives it to me again and asks I follow thru. Pressure was mounting, so on February 2, 2017, I finally give Space Command and the US Air Force what they'd been waiting for. It was at exactly 11:29pm that I sent them an email with the pictures - but no video. My story: the video file was too big for my system and I would have to find a way to do it. They would have to wait for it.

The very next day, on the morning of February 3, 2017, I got an email that Space Command is eager for my materials and that she hoped I had followed through. I replied that I already had. But only hours after that she frantically called me and said she got a cryptic message from her sibling that said only this: "Let Marian know that f-one-eights are coming his way because of what he saw". I asked her what this meant, and Sarah explained that her sibling simply somehow overheard this and didn't know what it meant but thought, based on the high level of her spouse's work, that I should know, and that hopefully I would understand."

I discussed this with my brother, and our consensus was that it meant F-18 jets. That's the only "f-one-eights" we could think of. Later that night as my mom was watching the 11pm CBS evening news I heard my mom literally screaming for me to hurry in! As I raced in I watched CBS report on the surprise arrival (according to them) of two F-18 Super Hornet fighter jets. Going on the internet revealed the report to be incomplete as sources clearly showed the presence of, at the very least, 6 Super Hornet fighter jets. My own observations from my house confirmed this. Close-ups by CBS were so good that they allowed me to even identify some of the pilots. Afterwards, a stream of F-18s circled our area, one after another, every hour on the hour. This went on for a week. The planes took special pains to fly all along the foothills above our house (where the crippled Craft-4 might have gone down).

According to news reports the planes were scheduled to leave at an undisclosed time on Sunday, February 12th. CBS News reported that they tried to reach out to NORAD and got no comment on why the planes were here. The only admission from the military was that these were Navy jets, and that they were here for an unscheduled (read "surprise") visit, and added no further details. Fortunately, with a little bit of detective work I was able to piece together a fairly accurate picture of what was going on.

There were a total of 7 (or more) planes. Six were F-18 fighter jets (and possibly more). My sources indicate that there were probably other "unseen airborne assets" that didn't use Bob Hope Burbank Airport for their base of operations (these most likely operated out of nearby Edwards Air Force Base). The seventh known aircraft was an Air Force E-11A Sentinel intel aircraft.

The F-18 Fighter Jets Arrived In 2 Groups
The first group was of four F-18 fighters from the "Fighting Redcocks" squadron (VFA-22) directly off the USS Nimitz aircraft carrier, which was probably sailing off the coast of the Los Angeles area. (The Redcocks home base is Naval Air Station (NAS) Lemoore, California.) They arrived mostly unnoticed by the majority of the public because they came during the day while the public was busy at work, etc. These planes were #105/NK, 110/NK, 267/NK, and a mysterious fourth unmarked/unnumbered plane with only the matching tail letters NK (as were all the planes).

Montage of images showing all 6 of the F-18 Super Hornet fighter jets on February 3, 2017. *(Multi-image montage; © 2017 Marian Rudnyk)*

The second squadron F-18 fighter jets, arrived later, in the evening, from their home in (NAS) Oceana AFB in Virginia. These planes were also assigned to Carrier Wing One. This second group consisted of two planes from the "Fighting Checkmates" squadron (VFA-211), led by Lt. Pete "Halfman" Harris and Lt. Zachary "Murderface" Sutherland, flying in the lead plane (#212/AB). The other plane was #262/AD. Their noisy arrival in the evening was what stirred news reports, and this was probably intentional so as to steer attention away from the larger F-18 contingent as well as any other aircraft that were part of the mission.

Also present was an Air Force E-11A BACN intel aircraft. It coordinated all these planes (along with "unseen" high altitude recon aircraft that may have included drones), including how they were all deployed and utilized. It also coordinated all communications and electronics assets. I mention drones here because later sightings by me regularly included military drones, thus strongly indicating that they were most likely a constant presence right from the start. The military, if nothing else, is always consistent.

Specifically, these fighter jets were "F/A-18F Super Hornets". Basically what this means is that these were top of the line supersonic, all-weather, capable of working from aircraft carriers, multi-role fighter-attack aircraft. They were also equipped with synthetic aperture ground mapping radar that allowed pilots to locate targets even in poor weather conditions. Additionally, they were night vision/night attack capable and used Hughes AN/AAR-50 thermal navigation pods and LORAL AN/AAS-28 NITE Hawk FLIR (Forward Looking Infrared Array) targeting pods, night vision goggles, and even 2 full color multi-function displays (MFDs; per plane) that included color map functionality. If that isn't enough, each plane was also fully reconnaissance capable by way of an ATARS electro-optical sensor package. All had added external fuel tanks for increased flying range.

It is important to note that the deployment of F/A-18F Super Hornets, into an "area of interest" is no small matter. These are more than just top-of-the-line fighter jets that the public thinks of as "Top Gun" planes. These are some of the most sophisticated fighting and surveillance machines ever produced on our planet. Add to that the presence of an E-11A intel plane and the plot thickens. This was an incredibly heavy-handed deadly reconnaissance-capable response to a UFO sighting – my sighting. But why?

Suddenly my sighting went from surrealistically astounding and seemingly unreal, to very brutally and dangerously real. Deadly real. The plane's mission was not disclosed. NORAD had no comment. The military called it simply "unscheduled training". The planes then patrolled the skies over my area for a whole week doing looping hourly flyovers in pairs. I was too stunned for words. They appeared to scour the nearby hillsides where "my" UFOs had last headed and disappeared.

Reaching Out For Some Answers…
When I discussed my sighting with Jane, she was not surprised by the planes appearance. Jane advised me that a common intelligence tactic was the these planes were the loud visible distraction – the so-called "shiny object" you stare at while the real work was being done elsewhere. I asked, "What do you mean?" She asked, "Did you hear any other planes around?" I replied, "Yeah, but I don't know planes by their sound (I have since gained experience here), and plus they were very high up, somewhere in the clouds probably, so who cares." Jane explained that often while the jets do their part, hidden higher up are recon and intel planes doing even more "work". These were very interesting observations for Jane to make, especially since I only told her about the two F-18s from the "Fighting Checkmates" (that arrived in the evening) and never mentioned the other four F-18s nor the E-11A BACN. Jane was more right than she knew… unless she really already did know…!? (A question I still can't answer to this day.) But most importantly is this validated everything I knew I saw and framed it in reasonable context.

It was then that I realized that there was one more piece to my puzzle that remained untouched, and now I had enough information to reach out to him – to the inimitable Bob Wood. Thus, on February 5, 2017, I finally called and spoke to Bob Wood. Our conversation was so amazing and fruitful that he even fired off a thank you email to me later that day (see "Appendix 4: Support Materials"). His insight into the military world and secret projects was extremely keen and invaluable.

Suddenly it all began to make sense! Thank you Jane, Stanton Friedman and Bob Wood! This was about more than just a UFO sighting. It was more than just about the fact that an expert observer, an astronomer no less, witnessed this event. And it was more than the fact that this expert witness (me) had both pictures and a video. It was about all these things combined with one very important key element: this was about possible crash retrieval!

4:

And Then The Real "Fun" Began

My Dreams Versus Fate: Santa Has A Sister, But The UFOs Don't Care
What do Christmas and UFOs have in common? Better yet, and more specifically: what do Christmas, Santa Claus, and Santa's sister have to do with UFOs? Not a whole lot at first glance. But Fate would step in and indelibly intertwine them all into a torturous mess – and massively complicate my life in majorly unforeseen ways.

So as 2017 wore on after my original January 1st sighting, everything became more and more clear to me in ways that are hard to explain. I reached out to both Stanton Friedman and Bob Wood once again, but this time was different. Each one had made it clear to me that they would not break my trust. On March 21, 2017 I spoke to Bob Wood, and on the 23rd to Stanton Friedman. I asked each to keep that confidentiality for me indefinitely until I said otherwise, and they understood and promised they would do so. They would only reveal my sighting to the world only if I gave them the green light. As of this book writing, now in 2019, I'm now giving the green light.

But back in 2017 things were different. Even though my "eyes were opened" so to speak, keeping my sighting "secret" was key because I had other plans. I had a Christmas book called "Santa's Sister", that I had published in December of 2015, so by January 2017 (the time of my sighting), that book was barely a year old. Thus, I had no intention in 2017, or in the future for that matter, of writing a book about my sighting. My focus was to be on solely making "Santa's Sister" succeed – and in the process, excitedly start on the super-Christmassy sequel I was eager to write and release as soon as possible. Space Command had asked me not to speak, so I explained this to both Stan and Bob, and they both agreed to stay silent until/unless I gave the word. My plan: I would shake myself free of all this UFO madness and go back to my original plans.

The only person now remaining was my brother. He was shocked and understandably angry that I had reached out to Stan and Bob, over his objections and warnings. And his warnings were dire. He had made it abundantly clear to me of the ruthless methods the government might employ against me if I talk. I explained in more detail everything that had happened and how both Stan and Bob had helped me see things more clearly, it was something I really needed, so he relented. Although he was not happy about it, he agreed that I did well and said that with Stan and Bob now in "quiet" mode, he would help me navigate the "UFO world" as any needs occurred. We also agreed to keep Jane "*in the loop*", so to speak, but to not mention Stan's or Bob's involvement lest it tip off "government forces". In hindsight this may have been a mistake, but it's hard to say even now, since even in 2019, things continue to evolve. In any case, my brother was adamant to stress that although technically I had done nothing wrong in any way, shape or form, that would not stop certain "sectors" of the government that dealt with the "UFO question" from seeking retribution or worse. At the time, I understood what he was saying, but still –in the back of my head- felt he was being a little over dramatic about all this. I told him I simply wanted to go back to working on my books, but he was adamant that I should write this book you are now reading. Nothing could be further from what I wanted to do. I wanted no part of this. He said that given everything that was happening, more things would happen, bad things, would happen and that Fate would give me no choice but to tell my UFO story. I shrugged it off. Unfortunately, for me, in the both the short and long run, he proved right – very right. So what do a Christmas book and UFO have in common? Like me, you're about to find out…

Sometimes You Move Back In Order To Move Forward
For me personally, all this UFO stuff was a moot point. Yes, I had seen UFOs, and had proof – big time – but you may be amazed to read, I STILL had no intention of going public. My brother was chomping at the bit to somehow eventually get my sighting out into the public spotlight, but I did not. My transformation, if you want to think of it that way, was not yet complete.

Knowing the truth, and doing something about it were two different things to me. I looked at it this way: it was only one sighting, so who cares. (And I had even put that second sighting of Jan. 20, 2017 totally out of my thoughts.) Lots of people saw UFOs. Yes, I was an astronomer who saw them, and that, I was told, was a big deal. But part of me still didn't care. There was possibly a crash. Maybe. There was a massive military response. Yes. Good. It means "they" (the government I pay taxes to, and to whom I therefore entrust my safety) took care of it, so in my mind: I was done. Thus, my plans were completely different. Even so, I had taken the time to quietly copyright my sighting in a 2017 document that was nothing more than just the sighting and its photography, as a way to protect myself. And that was as far as I was willing to go.

For me, I was focused on my Christmas book, "Santa's Sister", which I held dearly in my heart. The book had been a labor of love that took years to write. It represented my transformation from NASA astronomer, planetary scientist and Hollywood visual effects artist, into a full-blown book author. After years of people constantly berating and imploring me to not waste my time working making others rich, but instead utilize my many creative talents to further my own creative dreams – I did it! Unknown to everybody, "Santa's Sister" was actually my third book, behind an unpublished first book, and a second book (a collection of my poetry entitled "Bamboo Garden") that I had quietly published as sort of a *self-I-can-do-it* test. There were 64 more books that I had already half started that needed to be finished, plus 3 sequels to "Santa's Sister", so I was counting on being very happily busy cranking out book after book.

Discovering asteroids was a dream come true. Working as an astronomer using some of the premier telescopes in the world was also a dream come true. Being at NASA as a planetary scientist, making discoveries, and being part of planetary mission flight teams – also a huge dreams-come-true. Being on staff at Disney Feature Animation as an artist – also major dream come true. Working in Hollywood on blockbuster movies like "Titanic" as part of an Oscar winning visual effects team – icing on the cake!

I had done well. But I had big additional dreams. I wanted to see my own creative creations come to life – in print and on screens - and it was time to chase them. Unfortunately, life threw me a detour. My dad got two cancers and required ever-increasing levels of 24 hour care, and my mom was overwhelmed. My parents needed help – desperately. For me, family was always first, so I stepped in. I sold my house, put my things in storage, and moved back so that I could help my mom take care of my dad. Part of me thought this might even somehow, in some twisted way, help me too by allowing me to focus solely on writing in my down time while helping them. This proved to be a pipe dream. Helping my parents became a round the clock responsibility. My only recourse was to do my writing and drawing during whatever sleepless nights I could scrounge up.

I decided to choose a project, and it would be my new focus. I decided on "Santa's Sister", a Christmas story I had started in the mid-90s and had revived in a twisted cartoon version (in the early 2000s) that bore little resemblance to my original vision. Back then, in pitch sessions with the head of Creative at Cartoon Network – he privately implored me to dump this version and embrace my original vision. After considering this, as well as taking a hard look at my many other half-started books and projects, I decided that more than any other of my projects, "Santa's Sister" held most precious in my heart – so it would be the first thing I would try to tackle to re-launch this exciting new phase of my multi-faceted career.

As I dug in I decided to trash all my recent iterations of that story and go back to its original roots, as had been suggested to me. I masterfully started crafting an epic Christmas story, like none other. I wanted a Christmas story that spoke to the imagination and the heart – on a huge, yet still very personal scale – one that would allow people's imaginations to soar, while also touching their hearts.

The first seven years were agony. The writing was beyond tough, and all the while my dad's condition worsened. In 2011 my dad passed away and I was devastated. Now my mom needed help, so I stayed on, now helping her rebuild a new life and taking care of her. Somewhere along the way, by December 2015, I finished "Santa's Sister"! The book was released with zero fanfare and absolutely no promotion because at the time my mom was going in for heart surgery. She had wanted to read the finished book before she went to the hospital, so it took everything I had in me just to finish it for her, and release it on Amazon – all barely days before her surgery. There was no time for advertising, so I just published it and let the chips fall where they may, while I took care of her.

As my mom recovered and was able to come home for Christmas, we were pleasantly surprised to find that "Santa's Sister" had done not great, but ok for a totally un-promoted book. Things looked hopeful. I used 2016 to promote the book. The re-rolled out "Santa's Sister" had a brand new punched up and gorgeous color book cover, which I had created in response to near universal feedback that no one liked the original monochromatic, what I called "frosty", first version. The book, which also featured songs, now had recordings I also made as an additional promotion for this new roll out. However, at the insistence of my Hollywood and writer friends, I decided that the December 2016 re-release would be free, meaning that as a one-time only on-Amazon deal, the book would be available during that Christmas for absolutely free. The idea, my friends assured me, would be to promote the book by getting it into more people's hands and thus eventually reach a wider "paying" audience later.

They were right. As December 2016 closed, my free Christmas book peaked at an incredible #62 among all of Amazon's holiday offerings. Yes, I saw thousands of dollars of potential sales lost by doing a promo giveaway, but the potential for much bigger things in the immediate future loomed big. Among these wonderful possibilities was Hallmark's new promotion that in the summer of 2017 they would do a "Christmas In July" TV event where during the whole month of July, 24/7, they would air all their many Christmas movies, and even feature a premiere of a brand new one. This was truly a gift for me! That meant that I had a full six months to prepare a promo campaign to piggy-back off their event and sell my book, and then be able to sell it again that December. So rather than having to wait a whole year to sell, I could sell my Christmas book twice in 2017! What a lucky break! So as New Year's Eve, December 31st, 2016 approached, I was visibly excited – very excited! Things were beginning to turn around in big and wonderful ways! Little did I know, as I toasted in the New Year, that in only a matter of hours, all that would be shattered – smashed into a million pieces - by the arrival of UFOs the very next day, and I was to be the unwilling star witness…

At First It All Started Quietly Enough
So, of course, you already know what happened next. January 1, 2017 came and my sighting happened. I did my outreach to Stanton Friedman and Bob Wood. Talked with U.S. Air Force's Space Command, and watched as a heavy handed military response unfolded when top-of-the-line F/A-18F Super Hornet fighter jets swooped in, along with their intel assets. But was that all there was? Unfortunately, for me, it was all just an opening prelude of the madness to come.

One of Space Command's stated goals is "anomaly mitigation". As it turns out, that anomaly wasn't just UFOs but me, and I was about to be "mitigated" – or so they thought. It first it all started innocently enough with things you almost don't notice at first. Phone calls that mysteriously get cut off. Emails that never arrive to their destination, or the emails you don't get although your friends swear they are being honest with you when they say they sent you a reply. At first it was all very innocuous, and easy to dismiss. Cell phones are notorious for dropped calls. Same for home phones. Emails sometimes get "lost" in the electronic nether-realm of the internet. On and on. But this was different. It was an ever-escalating level of controlled interference and disruption. Start paying attention and you begin to notice the patterns, let alone the fact that it starts to get harder and harder to communicate in any way in the modern ways to which we are all so accustomed.

Within a matter of only a few months of my sighting, even the most mundane phone calls became nearly impossible to make or maintain. Eventually, there were even pop ups that would appear that literally said I was being "blocked". For the most part, during this time I simply treated this as simply a nuisance and worked around it as best I could. But people around me were getting noticeably frustrated. My mother, for example, reached a point where it became nearly impossible to hold a conversation on the home phone because either there was too much intentional electronic interference or the phone would simply cut off again and again and again. People began noticing that they all had no such problems, yet when they called us it was nearly impossible to deal with. Emails were even worse – they simply never went through. It got to the point that I even tested it and tried to send myself and email from one account to the other. It would never go through.

By June of 2017 it was abundantly clear that I had a very serious communications problem on my hands. After months of no contact my best friend, out of frustration, finally reached out to my brother. My brother explained what was happening, and my friend came over that we and we talked and cleared the air. As it turned out each of us had assumed the other had curtailed all contact for some reason, but as it turned out no email, text, voicemail message, nor anything else ever reached either one of us. Evidently it was someone's "diabolical" plan to estrange me from my friends. This scenario played out again and again with each of my friends. Fortunately all my friends and I are much closer than that, and whoever was doing this, failed miserably. Since then we have all resorted to encrypted communications to successfully stay connected, and I'll leave it at that.

Fortunately, as similar problems all escalated, I eventually found clever ways around them all. These workarounds weren't perfect, but they kept me semi functioning. What I didn't know was the horror that was still lurking just around the corner. Hampering communications would soon be the least of my concerns.

In the meantime I innocently went about my business, oblivious to what was about to happen. I continued my promotional preparations for the big *Christmas In July* push for "Santa's Sister". Hallmark was advertising this heavily and I wanted to be ready. To this end I had 12 social media accounts that were all going to work in concert in a massively coordinated social media campaign blitz. This also included a "social media deck" website through which I could seamlessly manage all my accounts.

It was mid-June when it happened. I logged on to my social media deck, at Hootsuite, only to suddenly find that everything was crashing. All my social media accounts were suddenly disconnecting from Hootsuite. If I didn't fix the problem soon I would have no "Christmas In July" promotional campaign left. It would destroy my financial plans in a incalculable way. Unfortunately, that is exactly what happened. I went directly to my social media accounts to see what was happening.

My Twitter accounts were first to go – all suspended. By the next day the same thing happened to all my Facebook accounts. Attempts to "recover" all my accounts via their so-called "procedures" all failed. I was being actively denied access and my pages all shut down and unusable. With no income, and no promo campaign to run, I decided to leverage the one thing I now had lots of: time. Whoever was a Facebook and Twitter couldn't be there all the time. If I reached out enough I should eventually reach someone who would act on my behalf in spite of any mandate against me. At least that was the theory. It was all I had.

I won't bore you with the details suffice it to say that I emailed relentlessly 24/7 every hour on the hour until finally there was a crack. After weeks of pounding emails at them, a tech at Twitter suddenly actually replied. He admitted that someone there had done this to me and that it was wrong. He issued me a formal apology and unlocked my account for me and suddenly I had one account back. That was July 12th. I leveraged that letter to push on all my other accounts and eventually each one gave me a reply and apology. The same was true for Facebook. By month's end I had all my social media back as if nothing had ever happened. But it was obviously too late. My beautifully and intricately designed campaign to promote my beloved Christmas book was in ruins.

Additionally, what raised the creep factor were the black unmarked cars, white unmarked cars, or even police cars that would follow us. A trip from home to the store and back consisted of a police car instantly appearing to either pass our house as we were getting into our car, or appearing behind us as we left home. After a few blocks the car would turn down a street, as if to appear disinterested, but that was just be a rouse because a second police car would appear, follow us for a few blocks, then turn down a street. You'd think that would be it, but you'd be wrong. After a few more blocks, the same thing and still another car. Once at the store a car would park and case our car, stakeout fashion. You can substitute unmarked black, or also white car, for this and you get exactly the same behavior. Anybody who's seen any B-movie spy show could see through this and notice it. It was so ridiculously obvious and amateurish as to rise to the level of laughable absurdity. In a twisted way you could argue that it was like we had an police/MIB escort service protecting us. They never directly engaged us, but they were ever present. This behavior persisted throughout 2017, and 2018. Evidently someone caught wind of the fact that doing surveillance of someone going to the grocery store or the gym every day was just that – a trip to the store and gym – and was, well, to be blunt: boring and uneventful (from a security perspective). Whatever they thought they could "see", by early 2019 these so-called "intel geniuses" finally realized there was no there there. So by 2019 a car(s) might still follow us - but only sporadically (as helicopters took over! But we'll get to that later.).

Of course, all this might also have served to intimidate us. Instead it quickly devolved into a game of let's see how far we can make them follow us – as we occasionally zigged and zagged through random neighborhoods just for the heck of it. But a part of me was angry: this was our tax dollars at work? Seriously?

Finally, on August 2, 2017 I reached out to Jane, who would only speak off-the-record, and Jane told me it's what's called a "denial of income" attack – and that I was being sent a "message". Jane explained that whether its police that follow me, or tech companies that hassle me, or whoever or whatever it is, all probably have no idea why they are doing what they are doing, just that they were instructed to do certain things in the name of "national security" and were not to question, just do. To prove it Jane simply said this: "Ask your neighbors". So in the next few days I did just that. The result? No one around was being harassed for anything. We were obviously being singled out. Our phone, internet problems, and other tech problems - no one else had any issues of any kind. None at all. And most certainly, no one was being followed.

As 2017 dragged on, so did these "behaviors" as they obvious sought to harass via intimidation of sorts. For me, I focused on re-inventing a new promo campaign for "Santa's Sister". This was still my focus, contrary to what any government "forces" may have thought. If they thought my focus was UFOs, they were wrong. But by November I had learned to have some "fun" with the various cars that tried to follow me. I would "game" them by popping the battery in-and-out of my flip cell phone (I did get a smartphone soon after, but that's another story to come). This would cause me to seem appear and disappear. I tested my theory several times during longer driving trips to writing seminars I attended, etc – worked like a charm. To their credit, though, they would always impress me with how they eventually would get wind of where I was.

On a side note, I'd like to comment that from this point on in this book, rather than constantly having to refer to "*mysterious government agents*", or whatever these various people were/are, I'm going to invoke the acronym MIB, for the sake of brevity, and use it as a generic reference to such people - rather than to have to constantly explain myself. It doesn't necessarily mean they are men, or dressed necessarily in black, just that they are some sort of "special/covert government agents". Where I need to add additional details I will. So now continuing…

MIB Goons Crash The Voyager Party
By November 16, 2017 I must have finally tried the patience of these so-called MIBs, because this time I would square off with them face-to-face in a meeting I would never forget. On that day, by invitation, I attended an AIAA (American Institute of Aeronautics and Astronautics, of which I am a member) event celebrating the 40th Anniversary of the Voyager Program. The NASA Voyager missions, as you may recall, were the wildly successful Grand Tour of the Outer Planets of spacecraft's Voyager 1 (to Jupiter and Saturn) and Voyager 2 (to Jupiter, Saturn, Uranus, and Neptune).

The event was being held at the famous S-Café at aerospace giant Northrop Grumman Aerospace Systems in Redondo Beach, California. Naturally, having been a member of the Voyager team during the Neptune Encounter, I was very excited! My former bosses, and NASA luminaries, Ed Stone (former Voyager Project Scientist, and later Director of NASA's JPL) and John Casani (former Voyager Project Manager) would be there, as would my best friend (unnamed here) who was a veteran of NASA's Galileo mission to Jupiter. It was a chance to reconnect with old NASA buddies – I couldn't wait!

It was a cold rainy night as I fired up my 1962 Thunderbird and headed out for the long traffic-clogged drive to this event, in style – and with MIBs in tow – of course. By now I simply assumed they must want me to know I was being followed, so I just went with it. About half way there I popped the battery out of my flip phone and was able to shake myself free of them. This was one time I simply wanted to enjoy myself without all this UFO-drama following me. Unfortunately, for me, the MIBs had other plans and would not be so easily deterred this time.

John Casani and me sharing a lighthearted moment at the AIAA Voyager 40th Anniversary event on November 16, 2017. (Personal photo collection, image DSCN3854, taken Nov. 16, 2017 at 22h03m26s with Coolpix S3100 camera; ©2017 Marian Rudnyk)

Arriving, I found that the event was packed. After the dinner portion, Stone and Casani took to the stage and each took turns reminiscing about their tenures on Voyager and eloquently fielded questions afterwards. Following this, there was a meet and greet mixer. Interestingly, except for Stone and Casani, I was the only other Voyager team member there. This, unexpectedly to me, made me somewhat of an attraction as well. The fact that I brought my old Voyager Neptune Encounter Mission Handbook added even more interest for the other attendees – who enjoyed flipping through its "analog" pages (as they humorously put it).

Ed Stone and John Casani were happy to see me. Ed and I reminisced about that day at JPL many years ago when, as JPL Director, he dropped by my office to see if the rumors were true that I really had an office that was nicer than his. To his shock – I did! We laughed and I added that it was only because I had managed to "recover" beautiful mid-century modern wooden furniture from the 50's and 60's, that I was able to give my office a rich retro-NASA vibe. He commented that he recalled my reputation for, what everyone liked to "affectionately" call, my "creative resourcefulness" (his way of nicely saying I had a knack for pillaging old NASA warehouses for cool forgotten old stuff). But hey, I saved money, and my facility looked great! You may recall me detailing this story much earlier in this book… it was nice to see that Ed had still remembered whose office was nicer! He definitely had, not only a good memory, but also a great sense of humor. Who says fellow scientists can't be funny…

Ed Stone and I, reminiscing and enjoying a laugh together. (Personal photo collection, image DSCN3848, taken Nov. 16, 2017 at 21h58m20s with Coolpix S3100 camera; ©2017 Marian Rudnyk)

Additionally, my best friend introduced me to someone new: famous science writer Rod Pyle, who he knew, and who had wonderfully moderated this event. As everything wound down we finally made our exit. It was truly a wonderful time – who would have thought that it would soon have a sinister twist and turn dark…

Seen here is the picture I took that night, at 11:33pm, of the portion of what was originally the TRW Space & Defense Park that doubled as the Deneva Colony in the classic 1960s Star Trek. (Image DSCN3862, taken Nov. 16, 2017 at 11h33m36s with Coolpix S3100 camera; ©2017 Marian Rudnyk)

The parking lot quickly emptied out (it was a Thursday, a workday after all), while my friend and I lingered and talked. We walked around a bit at Northrop Grumman. He wanted to show me some local TV show history. As it turned out, part of the current Northrop Grumman complex, he explained, remains virtually unchanged to this day. It was originally the TRW Space & Defense Park. That specific section, he explained, had been used for the filming an episode (on February 15, 1967) of the classic TV show "*Star Trek*". This was the episode called "*Operation - Annihilate!*", where Capt. Kirk and company beamed down to the Deneva Colony. I have to admit – being a huge Star Trek fan – that was indeed exciting. It was the prefect way to cap off what appeared to be a perfect evening… so far…

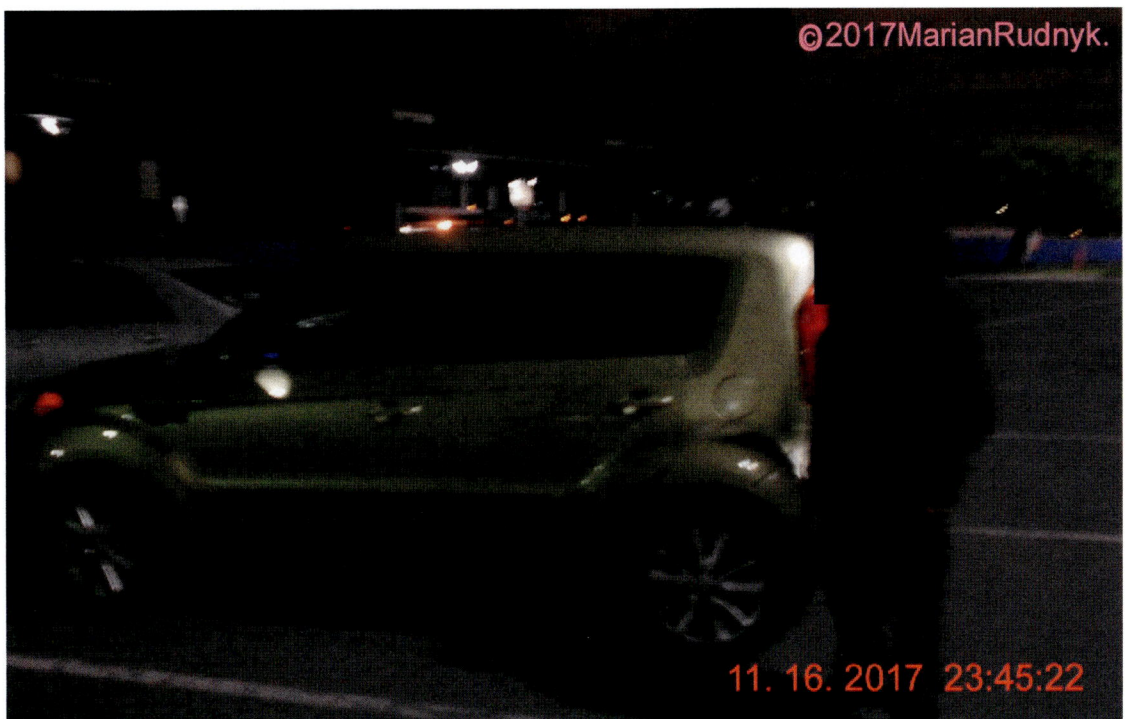

My friend (face redacted for security reasons), standing proudly in front of his new car, with fencing visible in the background. Although entry was "open" initially, the area was secured for the night, and a guard checkpoint controlled entry/exit by this point. (Image DSCN3863, taken Nov. 16, 2017 at 23h45m22s with Coolpix S3100 camera; ©2017 Marian Rudnyk)

As we got ready to depart, I snapped some pictures of my friend in front of his new car and then I got in my car, but it wouldn't start. My friend got out of his car and walked over and asked what was wrong? I replied that it was a mystery to me. The car was in perfect working condition, as was the battery. I do much of my own work on my cars, part of the joy I have of owning and driving only vintage cars, so I know well what condition they are in. These vintage cars are tough as nails, drive great, and look even better. My 1962 Thunderbird was my pride and joy and had no problems of any kind. I knew for a fact the battery was perfect. I was mystified. While it made no sense, the only thing that could be wrong was, well, the battery. I popped the hood, got out my battery-tester/booster box, and low and behold I was right – it was the battery. Unbelievably, it was deader than dead. It made no sense, but that's what it was.

As I hooked up my booster box, so as to begin charging my battery, a creepy feeling came over us. As we looked around we noticed an SUV not far off from us. Three men stood next to it. They weren't talking. They were just staring at, not my friend, not my retro-cool vintage car, but at me. I could sense who they were as the hair on the back of my neck stood up and a chill ran through me. The MIBs had found me. They hadn't liked being ditched. And evidently this time it was personal.

My friend now became nervous. He asked me if I was now ok? I told him not to worry, a quick charge and I would be on my way. And since my friend had a long drive home, I told him he should go ahead and head home. I told him I had my flip phone so no worries. If I had a problem I could always call AAA. He insisted I call and let him know once I was ok and on my way. I agreed, and he drove off. As he did the three MIBs got into their SUV, and drove up and parked closer to me. If there was any lingering doubts as to who they were, in that moment it totally and absolutely vaporized. When my dad had regaled me with stories of his days during World War 2 with the CIC (the US Army's Counter Intelligence Corps), he often detailed the ways you could spot these types of people. I knew the tells. I knew I had to be careful – very careful.

Suddenly they got out and appeared to lightly chat among themselves while watching me – but not in the friendly way you might expect friends to do. I smiled at them, and "played dumb". In the back of my mind I couldn't shake the feeling that somehow they were behind my car travail. It's possible that they simply turned on my headlights while I was inside at the event, drained my battery, then pushed the headlight knob back in – and I would be no wiser as to what happened. However they did it, the battery was now dead as a doornail.

As time dragged on, with my booster box connected all along to keep the charge building, I repeatedly tried to start my car, but somehow my battery was just far too drained. This really surprised me because I had a 56amp 1962 reproduction battery, it was very tough and hard to mess with, but evidently these MIBs found a way to do exactly that.

It was becoming evident to me that this was no accident. Then, in an act of pure insane bravery, or stupidity (you decide!), I simply approached them. They stood there – shocked - and suddenly silent. I happily introduced myself, and chatted about how great this Voyager anniversary event was, and how having been on the Voyager team, I had been excited to be here. They just stood there – silent throughout – staring. They must have thought I was out of my mind. (Good, I thought to myself). Time to double down… so I extended a hand and shook each of theirs and said it was nice to meet them and asked where they were from? Did they work at JPL, or NASA, or in aerospace? They looked at each other in apparent shock that I would have the audacity to interact with them. But I smiled and tried to exude as much confidence as I could possibly muster. In the past, Jane had warned me against interacting with these people. So in reality, I was in a very dangerous situation, and I knew it. The next move was theirs.

One of them (the one on the left), finally replied that he was from JPL. I asked what project, and he stammered and then named a program I had never heard of. The other one (the one on the far right) said he was from there too (of course). I stated that that was interesting, and that I had never heard of that program, and then turned my attention to the man in the middle. I asked him matter-of-factly, if he was from JPL too? He just glared at me and would not answer. For lack of a better word, he just looked utterly pissed. I figured he must be the one in charge. I guess my actions were not in his "intimidations manual".

Believe it or not I decided to go ahead and double-down even more (a dumb move, in hind sight), and so I asked again if he was from JPL? He still said nothing. At that point, I proceeded to just brush it off and then turned to the man on the left and asked if they could help me out. (Oh, you should have seen the looks on their faces!) I explained that obviously my booster box didn't have sufficient kick to start my car, and that I wanted to get home, and could they give me a jump? He replied, point blank, that he couldn't because he had no jumper cables. "No worries!" I replied with a beaming smile, "Cuz I do!" and as I briskly made my way to my trunk, I added, "Be right back with the cables." The MIBs, to put it bluntly, were mortified.

As I pulled the jumper cables out of my trunk, I glanced back at them from the corner of my eye and could see they still seemed utterly flabbergasted at my apparent boldness and didn't know what to make of me. (Good! I thought again to myself.) Evidently, they were supposed to intimidate me, but instead I appeared to them to be some kind of clueless science schmuck who had no idea of what they were trying to do to me. As I came back with the cables, one of them grudgingly got in their car and pulled it up to mine. As they popped the hood I connected the cars and told him to rev his engine a bit to give it some extra charging kick. With a scowl, he complied.

As their car charged mine I decided I was not going to let them off this easy, and every time they asked me to try my car, I told them, "Not yet, just a little bit more. I wanna make sure it starts good." At this point t was as if I was rubbing this in their face, and they looked furious, but complied again.

Finally, I tried to start the car. Amazingly it still wouldn't start. I was dumbfounded, but they appeared pleased. They quickly disconnected everything, and handed me my jumper cables. One of them simply uttered, "Too bad," to which I replied that it was no problem, I would just call AAA now. I shook their hands (to their shock) as I sincerely thanked them for their help, and then got into my car and got on the phone with AAA. They retreated back and then just continued to stand and stare at me. As I got off the phone with AAA I yelled out the window to them, "Good news, AAA is on their way." They just stared. I hooked my booster box back onto my car battery. It was fairly drained, but I figured whatever trickle of electricity it could deliver while I waited might help later once the AAA guy arrived. With AAA on the way, I banked that hopefully I had thrown them off enough that they would now not be able to take any physical action against me.

Unfortunately, even the arrival of AAA was weird. It took some time for him to get to me, and even then he got lost. As he closed in, I was on the phone with him, guiding him towards where the security gate was. I could see him across the massive empty parking lot driving along a parallel street. As he went out of sight and approached where the security gate was I could here him arguing.

He explained to the guards that he was there to help a stranded motorist. The guards told him to turn around and that there was no one here for him to help. Still on the phone, I immediately chided-in, imploring him to tell them to look across the parking lot at the 1962 T-Bird – that's the stranded car. He did so and they actually had the audacity to say that no such car was there. Fortunately for me, this guy was tough and wouldn't give up. He started yelling at them that he could see me from the guard shack and that if they didn't let him in he would call the authorities to find out what was going on. Thankfully - they let him in. Whatever had been planned for me, fortunately for me, was beginning to fall apart for the MIBs.

As he helped me out, he asked about the "goons" watching us, and I simply told him that evidently they were the "welcoming committee" because I didn't know who they were. As we worked the car, it finally started! I breathed a sigh of relief. As the AAA guy packed up I called my friend to tell him not to worry, AAA had helped me and I was about to head home. As I finally drove off, I waved at the three men by the SUV (my MIB "pals"), and thanked them for trying to help me. In familiar form, they just stared - speechless. I then thanked the AAA guy and headed home and he went on his way.

As I approached the freeway onramp I could see a car closely tailing me. More MIBs. I just couldn't fathom the reason for all this attention, but I had had enough. I popped the battery out of my flip phone (so they couldn't track it), and punched the accelerator! "Let's see if the can follow a '62 T-Bird with a massive V-8 and a 4-barrel carb!" I thought to myself. They couldn't. I easily lost them and disappeared in the traffic. As I drove home, I cranked on my '62 radio, set it to Radio Disney, and thoroughly enjoyed my rainy cruise home.

And Then Black Friday Turned Black
I make no apologies for my loose use of the term "MIB goons" throughout this book, because that is what they are. There may be those who may frown on my use of this phrase I've coined, but I stand behind it and will not waver. As you read my book I hope you take that term, not as some silly moniker, but with the ugly bitter utter disdain that I have for these people and those who work with them and support the so-called "work" they do. I'll boil it down to one simple thought: to harass and harm your own mostly patriotic citizens does not support the cause of national security, it only supports and aids the cause of aliens, or whatever/whoever it is that is flying around our skies with impunity and bothering our military and who almost certainly do not have our best interest at heart. That being said, as you read this section I think you will easily understand the source of bitterness. If these MIB goons think they can make my life miserable, then I sincerely hope this book becomes the total and complete bane of their existence, and Ultimately, aids those in power to take action and shut them down.
Ok, taking a deep breath and now continuing…

All through that November I prepared for my big Christmas promo campaign for my "Santa's Sister" book. First up would be a Black Friday sale. But it was never to be. The goons had other plans for me. Although my social media was now all working, Amazon and their KDP (Kindle eBook) site was not – at least not for me. I wasn't getting any sales, so I did some checking. I went on to the Amazon website and decided to search for my book. There was (and still is to this day) only one "Santa's Sister" and "Marian Rudnyk" on Amazon. Typing in my name got me nowhere. Amazon's built in spellcheck now suddenly kept kicking in and correcting my name from "Rudnyk" to "Rudnick" and then telling me that there was no such person. Frustrated, I tried "Santa's Sister". Nothing on the first page of results. So I kept clicking, on the next page, and the next page, and the next. After I while the results pages didn't even have Christmas products on them. This was insane, how could my book not be ahead of all these products in the search results.

Finally, on the 28th page of results, there was my book. But instead of happy, I was horrified. My book's listing was somewhere in the middle of that page. Above me were multiple identical listings for black baseball caps with the words, "Fxxx YOU" in bold white letters on them. Below my book were multiple listings for black T-shirts with the words "YOUR SCREWED" in bold white letters printed boldly in white across them. And, yes, the word "you're" was indeed misspelled as "your". Normally such a misspelling would make me laugh, but these search results were no laughing matter. No one was finding my book, and for good reason. Anyone who did find it was probably mortified by the products that were listed above and below my book. This was psyops of the first order, and I was literally the target. As I have mentioned before, my dad had told me about how these counterintelligence things worked, so to a certain extent I was not surprised, but just more appalled at the ugliness of the attack. No wonder I had no sales. I had to do something, and fast, or I would be financially buried yet again.

A quick call to Jane confirmed that this was probably no accident. How could it? It was so blatant, but I still wanted to hear it, as she called it: denial of income attack. These words were now haunting me – and costing me my livelihood. I immediately got to work contacting Amazon. As you might imagine, that initially went nowhere. For days I worked the phones and emails trying to get through to someone. Anyone. Amazon was a faceless cold computerized non-caring digital monstrosity. The few times I did get through, they told me that what I said was not true (even though I sent them real time screenshots and links) and either hung up or put me on hold and left me there to rot.

It was now the start of December, but I kept persisting. Eventually I got through to someone, and this time played it cool and couched my complaint in tech lingo that basically only hinted at a search engine problem. I explained that I had a Christmas book that was having tech issues, and with Christmas literally around the corner, it was a pressing issue that needed immediate attention and resolution. This style of attempt was not new. I had tried every angle imaginable and been totally shut down, but this time I got someone who was actually human enough to be receptive. I then got transferred from one person to the next. And finally I got someone who said they could help. Let's call her "FanGirl" (because that's how she signed her emails to me later on during the fixit process after she Googled my name and loved the things I had worked on in my career). FanGirl was a sort of rebel to say the least. I don't want to say too much so as to shield her identity, but I will say this much: even though she worked at a tech giant, privately she and her husband were off-the-gridders of sorts, and she became my guardian angel, and I will always be grateful to her.

FanGirl re-enacted my searches and much to her horror, got the same results (no surprise there). She then asked for my patience as she would look into this from the inside, and vowed to get back to me. She seemed sincere so I believed her and let her conduct her "investigation", or "poking around" as she liked to call it. The results were shocking to say the least. She called me to inform me that my account was indeed hacked – from inside Amazon! And it was done by some "DOD types", as she described them, who had pulled "strings" to do this to me. She said some had even come in person (I won't detail where) in order to ensure their "directives" (that's the term she said she heard they used) were carried out. (The MIB goons were at it again!) I could tell she was visibly upset by this. She told me she could undo this, but I needed to trust her. How she fixed everything is best left unsaid, but the fact remains, that it took time. She Ultimately, succeeded in helping me by mid December. That was the good news. The bad news was that I had lost most of my potential Christmas sales. It was just too little too late. Oh, I managed to still sell a few books between then and New Years, and even a scattered few in January. But that was it.

On the good side: to my pleasant surprise, during this time (December 2017) an interesting thing was happening at the same time: partial UFO "disclosure". On December 16, 2017 the New York Times outed the existence of the "Advanced Aerospace Threat Identification Program". This was an official government sanctioned $22 million dollar program that was studying UFOs. Additionally, two DOD military videos were also released (called "Gimbal" and "Nimitz FLIR-1"). Each showed F-18 Super Hornets in hot pursuit of a UFO referred to as a tic-tac because of its shape.

The timing of all this couldn't have been more coincidental if I had tried. The same day this all this news broke, my Amazon problem was fixed. Not only that, suddenly all my MIB problems went away. Like a flick of a switch. It was as if someone had turned a spigot off and all my UFO related problems suddenly stopped. I could now drive without being followed. Our phones came back to life, including both house phones and my smartphone. (Yes, I had finally upgraded from my flip phone.) All our cable TV weirdness stopped. And even emails and text messages were suddenly getting through. No static, no weird electronic sounds, no blocks, no nothing. I couldn't have asked for a nicer Christmas gift - save having my book sell too, but it was too late for it again. Among the good news of this "soft" disclosure, was one piece of bad news for me personally: the MIBs still managed to damage my book sales just before all the madness ended. But end it did, and the holidays, though penniless, were wonderful.

5:

Finally Some Answers

What's In A Name
Before we plunge ahead into the big revelations exposed by my sighting, I think that first a very quick word about terminology is in order. My main concern when writing this book is the utter lack of specificity of terminology concerning what are commonly called "UFOs". There is a total lack of consistency, and in some cases there are different, even competing, names for the same term.

I find the term UAP (Unidentified Aerial Phenomenon) to be wholly unsatisfying and confusing. On a technical level, I don't think that unknown craft flying here from other worlds qualify as a "phenomenon". Additionally, the word "aerial" in the term is misleading since these craft are what Luis Elizondo (former head of AATIP) correctly calls, trans-natural craft – craft – vehicles if you will – that can travel easily from space to air to water. The word "aerial", on the other hand specifically locks down the unknown craft as only being "airborne" in nature. Additionally, media as well as the public at large seem resistive to the term. I doubt that re-branding something, in an attempt to escape stigma, is going to work. Better instead to embrace "UFO" and then add some terms that can be "specifiers". To that end I will be coining to new terms for the first time ever: UFC, and the especially useful and memorable: "NTAC". Let me explain…

To this end I will be using the following terms to make things as clear as possible:

1. **UFO = Unidentified Flying Object**: Any and all objects, regardless of shape and type, that are genuinely "*unidentifiable*" and with "*no delineated or defined origin*". It's a still a great term, even the military still use it, and has universal understandability and makes sense.

2. **UFC = Unknown Flying Craft**: These are UFOs that now have been shown to be some sort of artificial (non-naturally occurring) "*craft*". An in-between term for something that falls between the cracks of a complete unknown (UFO), and the clearly identified NTAC (see next term). An alternate term that could be used here could also be "**IPC**" or **Intelligently Piloted Craft**.

3. **NTAC (pronounced "en – tack") = Non-Terran Aerospace Craft**: This is a very specific term that I am coining here for the very first time anywhere, and that I feel passionately can solve the whole naming crisis that seems to be presently occurring. **NTAC** defines exactly a flying object that has now been established to specifically be a "*craft*" – an intelligently controlled machine-vehicle of some sort. It also refers to craft that are established to be of an "other-worldly" origin – in other words alien or of an "*origin beyond our planet*". Thus a **Non-Terran Aerospace Craft,** or simply **NTAC**.

So we now can think of a UFO as being a nice general term for a true "unknown", while UFC, and especially NTAC, are nice terms that define *varying levels of identification*, NTAC being the most specific. With these well-defined terms in hand we can now proceed with nice clear discussions.

Now It All Makes Sense...!
After weighing all the facts at my disposal, including everything we've discussed so far, I finally had some basic answers. I now had three things I could be sure of:

1. The UFOs were NTACs

2. Craft-4 was damaged

3. Craft-4 might have later crashed, thus making this event "important".

How did I come to these conclusions? The logic is actually fairly straightforward and obvious – especially in light of all we have learned. Let's quickly run down the list again, but this time with analysis included.

1. The UFOs were NTACs (Non-Terran Aerospace Craft)
Firstly, it's easiest to simply eliminate what they are not, based on the qualities they exhibited. They definitely did not belong to us - to our military. They also did not come from any other country on earth. Why?

For starters, all 4 craft were silent (made no sound of any kind). (Note: I will be excluding Object-5 from further discussions, since it was only in the pictures, and I did not actually witness it directly). They had no flight surfaces (no wings, rudders, flaps, etc.). They showed no outward forms of propulsion (no engines, propellers, etc; and made none of the sounds associated with such systems). These properties mean they could absolutely not be any type of airplane, jet, helicopter, or drone (UAV; "Unmanned Aerial Vehicle"). There was also no exhaust of any kind, that would indicate chemical combustion.

All 4 craft flew perpendicular to the wind. The winds were strong and consistently gusting directly west-to-east, while all 4 craft were traveling directly south-to-north and were not affected by any atmospheric conditions. The winds were not only witnessed by me, but the photos clearly show flags in the area (including an extremely large over-sized one on a very high pole across the street) all fluttering in the wind west-to-east. The video also supports my personal observation as well as the photos, because you can clearly hear the constant clanging of the flagpole rope system of the McDiner flag constantly clanging in the background. All 4 craft flew with no signs of 'floating-drifting" motions, but instead flew in a determined controlled manner. Their flight movement made it abundantly clear that they were also not tethered together in any way, and no such ropes/tethers/strings were ever visible in any way. All these factors clearly and completely eliminate any type of balloon or balloon-like craft or "lighter-than-air" objects.

Therefore, we now know that no kind of craft exists in either the U.S., or any country, that meets these criteria. That leaves only one other possibility: that these were still Terran (Earthly) craft, but were "different" because they were some sort of U.S. experimental craft. However, this possibility is easily excluded because the Space Command contact clearly indicated surprise and was adamant that they were indeed "real" UFOs. Additionally, this is further strongly supported by the fact that if these were "ours", why would we send our best F-18 military fighter jets, and intel support aircraft, to "survey" the area a month after my sighting? Remember, my sighting was on January 1st, and the F-18 Super Hornet jets, and company, all came on February 3rd. If they were "ours", then "our" people would be aware of their own experimental craft being here. The only reason the planes would still show up here so late after the sighting, was that if the 4 craft I saw were indeed "not ours" - and because there was very strong possibility they might still find "something" (a topic we will cover in a minute in item #3 below). This was obviously the reason for the extremely heavy-handed military response.

By elimination that leaves only one possibility: these clearly are NTACs. The fact that military aircraft were dispatched to investigate makes a very clear statement: these NTAC are considered "threatening" in nature, because the military only responds to military matters. The strong possibility of recovery of such a craft obviously elevated the level of military response – a point Bob Wood made very clear in our discussions.

2. Craft-4 was damaged

As I've described earlier, the behavior of the craft was of flying craft that were somehow trying to aid an "injured" brethren. This is supported by the fact that they seemed to struggle to maintain their flight formation, and that the crippled craft was the last one. Also, historically speaking, lightning strikes are known of being capable of bringing down aircraft. Lightning strikes have even been documented hitting an Apollo craft during launch. So these are fairly common, especially in conditions on the day of my sighting – which included thunder and lightning. The fact that the craft descended out of a heavy cloud deck, in weather that had included rain and thunder, supports this notion.

However, it should be noted, that it is possible that Craft-4 suffered some sort of unexpected mechanical, or system, failure, that was unrelated to lightning. But to witness 4 such high tech craft flying, as I did, I find it very hard to believe that suddenly one of them would suffer a system failure - out of the blue. It's possible, but highly unlikely.

Lightning is therefore the most likely culprit. This particular conclusion represents a "mostly likely" scenario… for now. Therefore, it is my informed opinion that Craft-4 was indeed damaged – probably by lightning.

3. Craft-4 might have later crashed, thus making this event extra important.
I'm just going to go ahead and say it: my sighting was important because it was about a possible crash and retrieval. What I saw supports this. The powerful military response seemingly confirms this.

Later events (discussed in following chapters) demonstrate there may even be more at stake, but this one probable fact is at the core of everything that follows. Although Craft-4 crashing seems like an easy conclusion to come to, in reality it took some time. I had to reach out to a number of people to run through every detail in order to come to this decision. Without belaboring this point, probably the most evident item that sealed this decision was the obvious fact that military jets came to the site of the sighting and surveyed it for a week – a month after the sighting happened. As I stated earlier, if it had been one of their own x-craft they would have known about its flight, known that it had gone down, and recovered it immediately after my sighting. The only reason they had to show up once they got wind of my sighting, was because it wasn't theirs and they wanted to investigate it – and hopefully secretly recover the wreckage. Add to this the fact that my Space Command "contact" had, (probably inadvertently), blurted out that the "UFO is real" (quote, "…That sucker's real alright. Those are UFOs…") pretty much seals the deal. An aerospace craft of unknown, probably non-terrestrial origin, was a big deal, and that is why there was such a swift (after they found out from me) and heavy-handed response.

So did they actually retrieve the craft? I don't know. As disclosure happens perhaps it will eventually be revealed. But as future events unfolded later in 2018 and early 2019, you will see that it becomes quite obvious that there is even more at stake.

Now Putting It All Together - The BIG Reveal!
Interestingly, there were 2 other UFO sightings that same day that were posted on YouTube. One before my sighting, and one after. This is important because it shows that my sighting was not a solitary event that day. Something was indeed going on that day.

The first one was at 12:09am (about 16 hours before my sighting) that happened in San Jose, California (roughly 349 miles north of me). It showed 3 UFOs flying in formation similar to my sighting (minus one). Later that night, at 1:37am someone saw 3 UFOs flying in the same formation in Reseda, California (only about 39 miles from me).

One could speculate that the 3 craft were coming to the aid of their stricken friend who was in my area. They were somehow informed of their "friend's" plight and raced down from the San Jose area, and began searching for him, until they finally found him struggling in the clouds above Monrovia, that afternoon. They were able to partially help this stricken craft (so that it wouldn't literally crash in downtown Monrovia – you can only imagine the sensation that would have caused!), but in the process all four dropped out of the clouds for a few precious minutes, and as luck would have it, I was there and saw this. They managed to disappear into the cloud-covered mountains, but something went wrong (perhaps Craft-4 was simply far too damaged), and the struggling craft crashed. The three assisting craft were forced, after possibly rescuing any survivors, to abandon their stricken friend.

Why would they do this? The most obvious and easiest explanation is that as night approached, the local situation here, changed dramatically. Remember, that although it was January 1st that day, there were no New Year's celebrations because it was a Sunday, and as per tradition here, everything was bumped to the next day (January 2nd). Because of this, activity was ramping up - including military activity as January 2nd approached.

But why would the military be a concern to them? One reason was the B-2 Spirit Stealth Bomber was going to do an over-fly of the Rose Parade early that morning, and then later again over the Rose Bowl game that afternoon during its opening ceremony. If that wasn't enough, the opening ceremony of that Rose Bowl game honored World War 2 veterans, so there was a huge outpouring of military people present at the game. So the military presence increased dramatically. Additionally, both the parade and game featured aerial coverage, including a blimp.

The weather was also a problem because the day was cold and cloudy, with scattered rains in the area. However, the fact that the sky was heavily overcast was also a blessing. If they were to escape, it would provide cover.

Also, it's even possible that they were aware of the fact that I saw them. That I don't know, but it is a possibility that can't be ruled out, since they did fly fairly low, and I wasn't hiding when I took my pictures and video. By the morning of January 2nd they would risk many people like me possibly photographing them. One witness may not have represented a problem for them (as far as I can reckon), but thousands lining a parade route would definitely not be a good thing for them.

This meant that with each passing minute, as the morning of January 2nd approached, the aliens, if they were still trying to help the stricken craft (Craft-4), were forced to make a decision. They obviously couldn't stay. With each passing minute, the longer they lingered, they risked exposure. So they couldn't stay, and the clock was ticking.

Additionally, if there were survivors of the crash, they may have needed treatment. Ultimately, I feel it becomes quite obvious that they had to cut their losses and make the hard choice: rescue their fellow aliens, carefully destroy what they could of the downed craft, and then leave – while they still could.

Even completely destroying the craft may not have been possible because it would draw attention (smoke, fire, etc.). They probably did the best they could – which probably included destroying as much technology as they reasonably could, and then leaving. As they headed off, it was only hours before the Rose Parade. Also, locally activity was quickly ramping up – both military, general public, and national and international media.

As they sped off, at around 1:37am they were spotted over Reseda (heading north-northwest, in the direction from which they originally came), and then managed to evade any further detection after that.

As January wore on, there were lots of rainy days, including heavy rains, and even thunderstorms. The UFOs either didn't have a reason to come back, or couldn't for a variety of reasons. In the meantime, as I mentioned before, on January 15th circumstances conspired against me and forced me to reveal my sighting to U.S. Air Force Space Command. Because of this, the military response had probably already been partially planned. Thus, once I handed over my photography to them on February 2nd, the very next day (now having photographic confirmation firmly in hand), they immediately dispatched F-18s to the area, along with other surveillance, recon, intelligence, and possibly recovery aircraft. If ever the "aliens" were hoping to come back and retrieve what was left of Craft-4, all this made it nearly impossible – at least without risking detection – by the military, no less.

As for the mysterious fifth craft that was found in my pictures, I don't know how it fits into all of this. Its presence remains a mystery to me to this day. It's a mystery which perhaps somehow time will solve – or that perhaps the military has already unraveled.

Now although this scenario is speculative, the sightings are all real, and the facts all strongly appear to fit what we know so far – or at the very least, closely resembles the actual truth. The reality of all this is that there is further evidence that indicates that all this may not be as speculative as it may first appear right now. Further revelations, the result of the stunning events that occurred in 2018 and 2019 are both eye opening, and shocking. All these events were just the tip of the iceberg. So Unfortunately, for me, this story still didn't end here. Indeed, 2017 has turned out to only be the very beginning.

6:
Some Final Thoughts

For the start of my closing thoughts I want to take a moment to mention my dad. You see, as overwhelming as all this UFO stuff has been, surprisingly there was one group of people that I was actually semi-prepared to handle. And it was my dad, when he was still alive, who had unknowingly prepared me for them: the MIB goons. Also, I want to be really clear here: coining the seemingly catchy term "MIB goons" is not meant as a joke. I use this as a serious and bitter term. And that is because, all kidding aside, these so-called MIBs, as you've now seen, are indeed no joke. I am now convinced that during the incident that transpired at the Voyager event (that I revealed earlier in this book), that I could have indeed come to serious harm or worse. That I am here to now write about it is because of my dad.

So what had my dad done to help prepare me? The best way to understand this is by briefly explaining exactly *who* my dad was.

The man everybody here in the US knew as *my dad* was not the same person he used to be. Here, stateside, he was an aerospace engineer who worked his way up the military industrial complex ladder to Ultimately, do classified military aerospace work for such elite companies as Astro Science Corporation, Conrac, and many others including the Datatape Division of Kodak. His work included interesting classified things like highly classified military spy satellites (like the BlueJay-D) and working on the Apollo moon program. He was a key engineer on the communications systems for the LEM, the Lunar Excursion Module, which is NASA-speak for the craft that landed our astronauts on the moon. Later, he also worked on engineering-recording systems for NASA's Space Shuttles, as well numerous US military space projects.

But there was another side. Before aerospace, in his past, he was a survivor of Soviet communism in Eastern Europe. During World War II he also endured Nazi oppression, escaped from Auschwitz, and somehow survived a Nazi slave labor camp, which he also ultimately, escaped. As the violence of World War II swept through Europe, my dad leveraged his advanced language skills (he was fluent in five languages from an early age) to the advancing American forces and was recruited to work as one of the youngest operatives for the US Army's elite CIC (Counter Intelligence Corps).

After the war his CIC work continued in concert with Allied post-war agencies such as UNRRA (United Nations Relief and Rehabilitation Administration) where he worked undercover ferreting out runaway Nazi war criminals, Soviet defectors, Soviet moles, and various spies. Later, reuniting with the surviving remnants of his dislocated family (over half perished in Nazi and Soviet camps), he immigrated to the US where he made the unlikely transition from concert violinist to aerospace engineer.

My dad in the 1940's. As you can see by those cold steely eyes, this would not be a CIC agent you would want to tangle with. (Author's Personal Photo Collection; ©2019 Marian Rudnyk)

So besides being obviously proud of my dad, why did I go over his background? Because his extensive knowledge of counterintel and psyops was something he shared with me. And he was so good at what he did, that to his last days, few people knew of his intel training. He made me deeply aware of the tactics used and how such government organizations worked at a level most people rarely experience, or have access to. It was this knowledge that now has helped me. As you saw, as my narrative unfolded, it has helped guide me through my entry into the twisted dark underbelly of the world of UFOs. Men-in-black and Project Blue Book… it is a world, like most people, that I was somewhat only superficially aware of, but had no idea how deep or dark it ran. Nor did I ever expect to be plunged in the middle of it. His words have helped me keep a cool head, stay disciplined, and focused. It also meant I knew what to expect, and how to deal with it. Thanks dad…

That being said, I need to make something absolutely crystal clear, my story is far from over. As you read this book, already events are unfolding on an ongoing basis that defy any semblance of reason to me. I am seeing and documenting things that I could not have imagined. Nor was it ever my intention to be a part of any of this.

And Space Command, if any of you are reading this: I think that in general you do really good work, but if my words are rattling you and your military brethren, please know that you only have yourselves to blame. You created this intersect in my life. All you had to do was leave me alone and I would have happily and unknowingly gone on with my life. But instead you somehow set things in motion that put me on a collision course with all the things that you and your ilk have been so keen to conceal. Now that cat is out of the bag. Time for you to step up and tell the public the truth. Unfortunately, I doubt that you and all your various intel associates have the guts to do what's right. But maybe someday soon, as events unfold, you will. I guess hope truly does spring eternal. My sincere hope is that you will eventually (hopefully soon) do the right thing and protect us by actually *protecting us*, and not hurt those you serve – which, by the way, *is* us.

So as I mentioned, there are ongoing events unfolding even now as I write this book. As a matter of fact, things are evolving so rapidly, and in so many directions, and have had increasingly growing and tremendous pressure placed on me to release what I know and speak up – especially about this first original January 1, 2017 sighting. As it turns out, the events I have told you about in this book have turned out to be much more important than I could have anticipated and ever imagined, and have caused a number of unexpected things to evolve. This forced me to divide the information I have into digestible understandable parts. Thus, this book is just the first part in an ongoing and evolving story. Personally, I would preferred to have been done with all this madness, and simply move on, but even I now have realized how extremely important it is for me to continue on and keep sharing until things come to some sort of reasonable resolution.

So let me take a minute to share with you a few tidbits of what has been happening and will be in my upcoming sequel, "**Intersect 2: The Game Changers**". Some of the things featured will be not only the utterly over-the-top hostile actions of the helicopters around here, but two absolutely critical things: off-the-scale numbers of jaw-dropping UFO sightings and the massive military response that goes virtually unnoticed – even as actual battlefield warplanes fly overhead!

Let me give you just a tiny taste of what's been going on…

On February 7 of this year three Hawkeye intel planes circled over Monrovia – I repeat: Monrovia, CA. As bizarre as this may sound, even stranger is that fact that the planes were being followed by two UFOs. I saw it. It was incredible… and I caught it all.

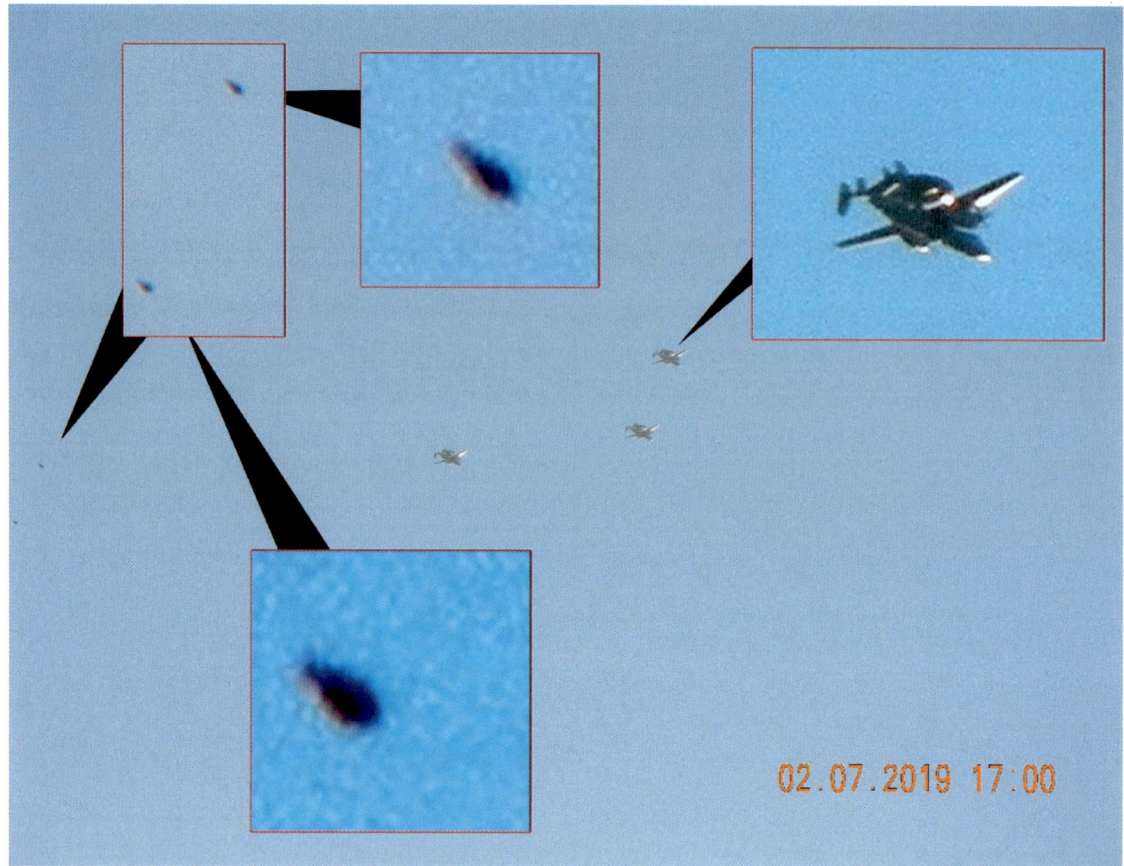

On the left we see two UFOs suddenly drop out of the sky and follow behind a trio (!) of Hawkeye battlefield intelligence aircraft. (Montaged version of image DSCN1572 taken Fe. 7, 2019 at 17h00m06s; ©2019 Marian Rudnyk)

This begs three really obvious questions. What are war craft (especially three of them!) doing flying over Monrovia? Why are they being followed by UFOs? And what are UFOs doing here in the first place?

Speaking of being followed, what about this DA40 spy plane and the tic-tac UFO that is following it?

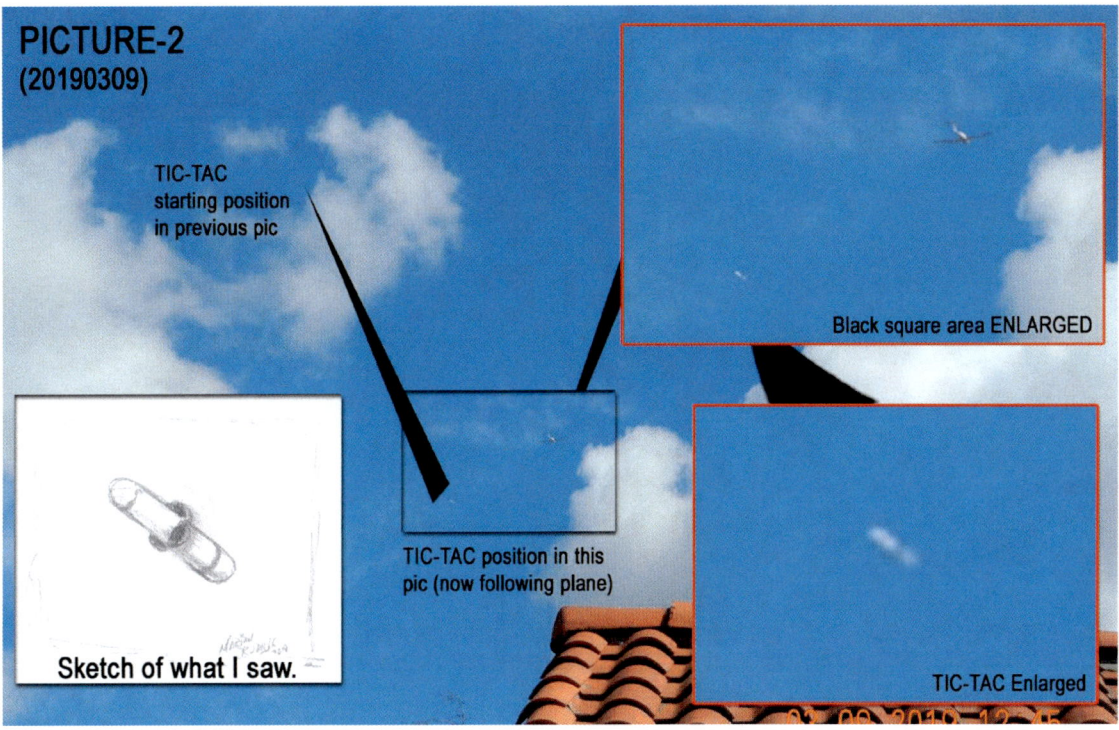

Look closely on the lower left corner, of the picture at left – yes, that is indeed a tic-tac UFO behind that plane. This tic-tac was a white capsule shape with a central dark "lumpy" ring. Seems for a spy plane, the DA40 doesn't "see" what's behind it very well. (Image DSCN3043 taken March 9, 2019 at 12h45m44s; ©2019 Marian Rudnyk)

And if this image wasn't so worrisome, then it would actually be kind of cool…

The picture on the left was taken this last January and features a strange triangular UFO I caught openly darting about in the early evening sky. (Image DSCN0752 taken Jan. 30, 2019 at 17h49m26s; ©2019 Marian Rudnyk)

And then there's those "pesky" MIBs who are still at it. They escalated things to dangerous levels (even sabotaged my car) and then went ahead and upped the ante off the scale when they did this on March 28th of this year:

This military woman, with the regulation hair and all, parked herself right in front of our house and started shooting photos with this massive 35mm ultra high powered digital telephoto rig – and then surveilled the house with some sort of weird scanner. All caught in pictures (left) and video. And yes, I got her face too! (Image DSCN3599 taken March 28, 2019, 9h26m50s; ©2019 Marian Rudnyk)

And then there were more than an air show's worth of battle-ready helicopters here. Everything from V-22 Osprey, Apache's, Black Hawks, and even Vipers – just to name a few. I got pictures and videos of them all! And not only were they flying over the house, but sometimes they circled, or even took "runs" at me from the air.

This armed Viper circled our house, until he realized I was filming him, and the he took off. But why was a serous battlefield AH-1 Viper here in the first place? And he's just one of many that have come since 2018. And they are secretly landing behind our hills too! What is going on? (Image DSCN3618 taken on March 28, 2019 at 13h12m56s; ©2019 Marian Rudnyk)

I love V-22 Ospreys, I think they are amazing aircraft – for a war zone. So why are they making continuous appearances in my area? And even over our house! Why are they unmarked? And why are they landing behind the foothills above our house?

This lone unmarked Osprey (at left) circled our house with the rear door open. Standing inside were men with some sort of equipment that they were pointing at me. Why? What was it? And who exactly are they? (Image DSCN2696 taken on Feb. 28, 2019 at 13h17m22s; ©2019 Marian Rudnyk)

Same goes for a whole parade of unmarked MIB, as well as military, helicopters such as the one below…

This helicopter, took a serious vicious aerial dive "run" at me and then shot off and landed in the canyon behind our foothills, on April 22, 2019 - as do many other helicopters of all models. What's back there – a secret military air base?! If so, why? (Screen capture images of video DSCN5284 taken April 22, 2019 at 13h46m18s; ©2019 Marian Rudnyk)

And why are we inundated with Black Hawk battlefield helicopters here? And if you try to take their picture, beware, they will try and take a "run" at you too. Why? What are they doing here?

This Black Hawk took a run at me that was so low, just look at the good picture I was easily able to snap. Why the interest in me? And why/how was it "stationed" in the canyons behind our hills? Because that's where it came from! And what is it doing here? (Image DSCN2734 taken March 4, 2019; ©2019 Marian Rudnyk)

And speaking of taking aerial diving runs at civilians, take a look at what this one did to me. And this has become more common than you might imagine. Why?

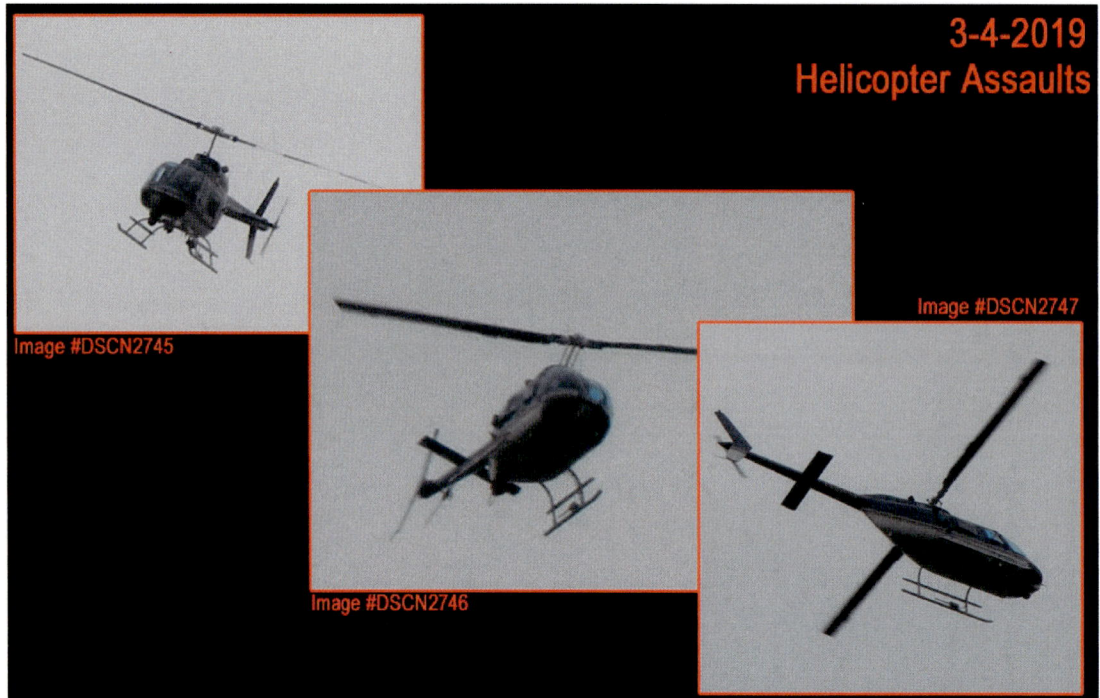

This helicopter not only took repeated dives at me, as did another he worked with. Shockingly 2 Black Hawks hovered high overhead and were coordinating this vicious harassment. But why? Since when is this ok? (Images DSCN2745, 46, & 47 all taken March 4, 2019; ©2019 Marian Rudnyk)

Perhaps besides me, they're also interested in these…because white orbs, such as this one, are now actually common here – why?

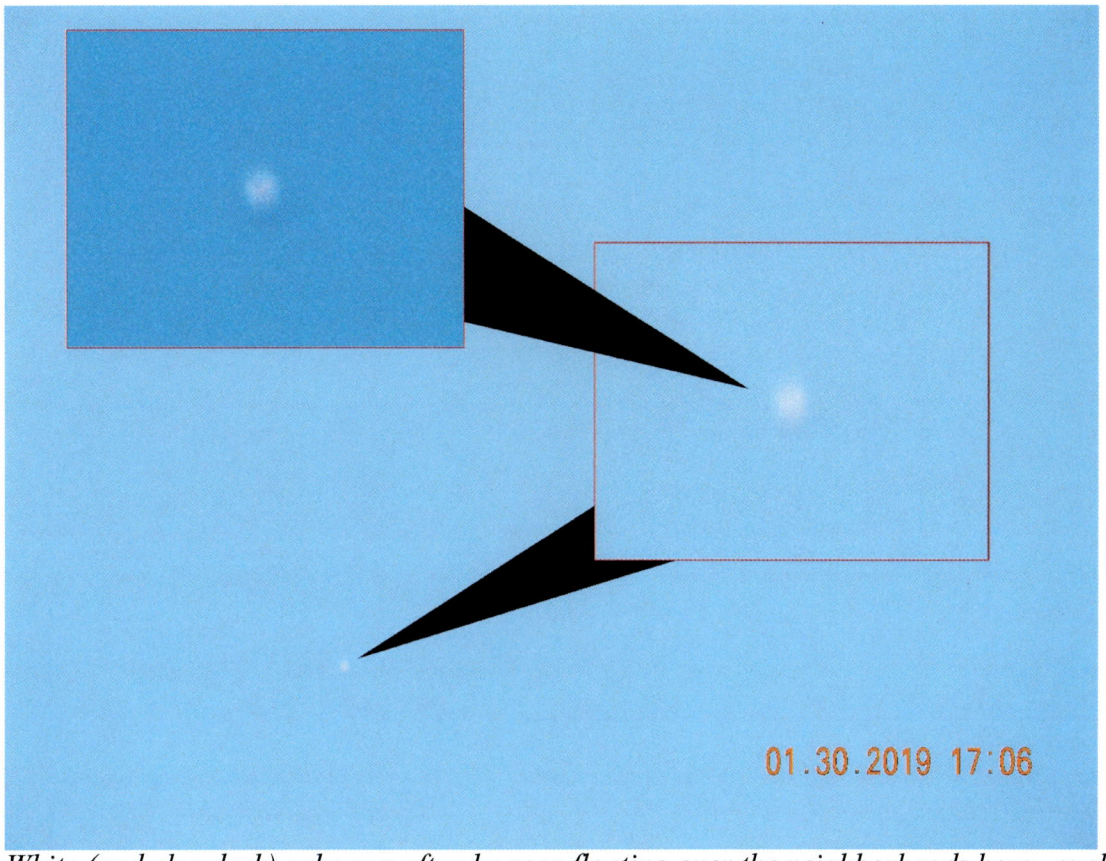

White (and also dark) orbs can often be seen floating over the neighborhoods here - such as this one on Jan. 30th this year. Then in phenomenal bursts of speed, they disappear into the sky or clouds. Some simply follow behind unsuspecting airplanes and jets. Why? (Image DSCN0688 taken January 30, 2019, 17h06m30s; ©2019 Marian Rudnyk)

Seems the military aren't pleased, though they're not talking. But evidently they have no problem working with unmarked MIB helicopters in trying to chase these objects.

This helicopter was part of a group of helicopters that took turns circling this white orb that hovered over our neighborhood and refused to leave. (Screen capture images from video DSCN4532 taken April 8, 2019, 17h06m30s; ©2019 Marian Rudnyk)

And a "new" never-before-seen (to my best knowledge) UFO makes its debut: an incredible looking pyramidal tetrahedron…

VIEW:
Facing S, ~60deg above horizon, at Foothill Gym parking lot, Monrovia, CA; note the strange striped clouds, & contrail.

CLOSE-UP OF:
02.28.2019 13:56
[at 2:56:30pm]

This strange "new type" UFO has been making regular appearances, including hovering directly over my house. What exactly is it? Why is it here? And why the interest in this area, or for that matter, me?! (Image DSCN2721 taken Feb. 28, 2019 at 14h56m30s; ©2019 Marian Rudnyk)

Lastly, why are there military drones regularly patrolling the skies of southern California? And that includes this new secret twin-engine military drone that isn't supposed to even exist, but evidently does…

A new secret twin-engine military drone (at left) makes its debut over the skies of Monrovia and southern California January 19 this year. It's easily identifiable by its short corkscrew like contrail. (Image DSCN0248 taken January 19, 2019 at 17h19m34s; ©2019 Marian Rudnyk)

And that was only a small part of a much uglier and stranger narrative that has unfolded. Is it possible that UFO activity is so intense here that the military has set up an actual secret military "forward" airbase within throwing distance of unsuspecting residential neighborhoods? And why all the UFOs? Has this area literally become what I call a "UFO transit zone"? Why? And why does our military seem powerless to do something about this?

All this and so very much more, all packed into the groundbreaking sequel to this book: **"INTERSECT 2: The Game Changers"**, coming soon during the winter of 2019!

So as I close here I am reminded of my conversations with Stanton Friedman. Stan had very much personally lamented to me the irony, as he put it that: he spent a lifetime studying UFOs, and yet had never managed to actually seen one himself. It struck him with sheer irony that this was the case, and yet a guy like me (nothing personal, he would add) could simply step outside of a McDonald's and see not one, but four UFOs. In hindsight, I wished I had invited him to visit Monrovia –his dream would have easily come true.

APPENDICES

APPENDIX 1:
Photo Archive

In this section, I archive all the pictures that were taken January 1, 2017 during my main first sighting. For the sake of completeness, I include all the pictures that were taken during the event, defined as the time I arrived at the McDiner to the last picture taken before I left. Note that each frame name (file name) is preceded with "DSCN" followed by a four digit number (#) and then the extension ".JPG". This results in the format DSCN####.JPG. For the sake of clarity, the "DSCN" and the ".JPG" have been dropped. Thus, image file "DSCN1657.JPG" becomes simply "1657". Frames with known objects are bolded. In future editions I hope to further expound on some of the amazing imagery from the other sightings I had after this first one. (All images ©2017 Marian Rudnyk)

SPECIFICATIONS:

All pictures in this section were taken: Sunday, January 1, 2017
Camera used: Nikon Coolpix S3100 (the only camera I had with me by chance)
All times are: Pacific Daylight Savings Time
All images are: 4320 x 3240 & with a Bit Depth of 24
Resolution (Horizontal & Vertical): 300 dpi
Color Representation: sRGB
Focal Length: 5 mm, unless otherwise noted in "Comments".

IMAGES:							
#	Name	Size (KB)	Picture Taken	F-#	Exposure (sec.)	ISO Speed	Comments
1.	1657	5932	2:47:26pm	F/3.2	1/640	ISO-80	A
2.	1658	5864	2:47:50pm	F/3.2	1/500	ISO-80	A
3.	1659	5764	2:47:58pm	F/3.2	1/500	ISO-80	A
4.	1660	5829	2:48:10pm	F/3.2	1/640	ISO-80	A
5.	1661	5745	2:48:16pm	F/3.2	1/1,000	ISO-80	A
6.	1662	5766	2:48:28pm	F/3.2	1/400	ISO-80	A
7.	1663	5777	2:48:38pm	F/3.2	1/500	ISO-80	A
8.	1664	5879	2:48:46pm	F/3.2	1/500	ISO-80	A
9.	1665	5896	2:48:54pm	F/3.2	1/500	ISO-80	A
10.	1666	5797	2:49:08pm	F/3.2	1/400	ISO-80	A
11.	1667	5790	2:49:20pm	F/3.2	1/320	ISO-80	A
12.	1668	5853	2:49:32pm	F/3.2	1/400	ISO-80	A
13.	1669	5812	2:49:40pm	F/3.2	1/500	ISO-80	A
14.	1670	5819	2:49:52pm	F/3.2	1/400	ISO-80	A
15.	1671	5802	2:50:04pm	F/3.2	1/800	ISO-80	A

16. 1672	5848	2:50:18pm	F/3.2	1/50	ISO-80	mom in car
17. 1673	6036	2:58:04pm	F/3.2	1/30	ISO-80	A
18. 1674	5833	2:58:16pm	F/3.2	1/40	ISO-80	A
19. 1675	5663	2:58:32pm	F/3.2	1/30	ISO-80	A, me & mom
20. 1676	6002	2:58:44pm	F/3.2	1/30	ISO-80	A, me & mom
21. 1677	**5867**	**4:22:58pm**	**F/3.2**	**1/400**	**ISO-80**	**A**
22. 1679	**5853**	**4:23:52pm**	**F/3.2**	**1/40**	**ISO-80**	**C, F**
23. 1680	**6174**	**4:24:00pm**	**F/6.5**	**1/50**	**ISO-80**	**C, F**
24. 1681	**6404**	**4:24:12pm**	**F/5.8**	**1/60**	**ISO-80**	**C, E**
25. 1682	6578	4:24:28pm	F/5.8	1/800	ISO-80	C, E
26. 1683	6364	4:24:36pm	F/5.8	1/30	ISO-80	C, E

NOTE: #1678 is not listed here because it is the video file.

Comments Key: A = Photo of McDiner exterior (outside)
                  B = Photo of McDiner interior (inside)
                  C = Street view & clouds + Unknown Objects
                  D = Landscape/city view
                  E = focal length 14 mm
                  F = focal length 23 mm

#1657   2:47:26pm

#1658   2:47:50pm

#1659   2:47:58pm

#1660   2:48:10pm

#1661   2:48:16pm

#1662   2:48:28pm

#1663   2:48:38pm

#1664   2:48:46pm

#1665   2:48:54pm

#1666   2:49:08pm

#1667　　　　　　　　　　2:49:20pm

#1668　　　　　　　　　　2:49:32pm

#1669　　　　　　　　　　2:49:40pm

#1670　　　　　　　　　　2:49:52pm

#1671　　　　　　　　　　2:50:04pm

#1672　　　　　　　　　　2:50:18pm

#1673   2:58:04pm

#1674   2:58:16pm

#1675   2:58:32pm

#1676   2:58:44pm

#1677   4:22:58pm

#1679   4:23:52pm

#1680　　　　　　**4:24:00pm**

#1681　　　　　　**4:24:12pm**

#1682　　　　　　**4:24:28pm**

#1683　　　　　　**4:24:36pm**

APPENDIX 2:
Video Screenshot Gallery

This section contains an archive of screen capture images from the one video I shot on January 1, 2017 during my first original sighting. They are selected screen capture images from the video (DCSN1678.AVI, aka #1678), with the objects marked.

The video lasts about 34 seconds. These images are meant to be, among other things, a reference guide. Each image is tagged with "T = x", where x represents approximately how far into the video the image was captured. It is important to note that the video was zoomed in and out and there was also camera movement as I followed the objects - so size and perspective change. Since the objects go behind the various light posts at various times, it might be possible to tie down more precise speeds based on these occultations.

This set is not complete (i.e. not second-by-second), but is made up of selected representative images that I could obtain with the limited computer/software I have.

Unfortunately, when started filming the video focus-tracking squares (5 of them) appeared and began literally "dancing" all over the viewscreen. The result was that the craft lost their detail and became dark fuzzy dots that pulsed as they moved. In reality, as I observed the craft, obviously no such "pulsing" was observed, thus indicating that the camera was somehow being interfered with by the craft. The result is the craft become intermttently visible/invisble in the camera footage.

The camera has never done this before or since, thus indicating that proximity to these craft somehow interfered with the camera - because the video sequence represents the closest point the craft were to me. Even so, the video should be seen to be truly and fully appreciated – especially since Space Command were so extremely keen to have a copy. It still contains a wealth of information – especially audio (gusty wind, silent craft, etc.). The video is available online on YouTube, my Amazon Author Page and at www.Rudnyk.com. (All images ©2017 Marian Rudnyk)

T=0

T=1

T = 3

T = 3; close-up and enhanced.

T = 4

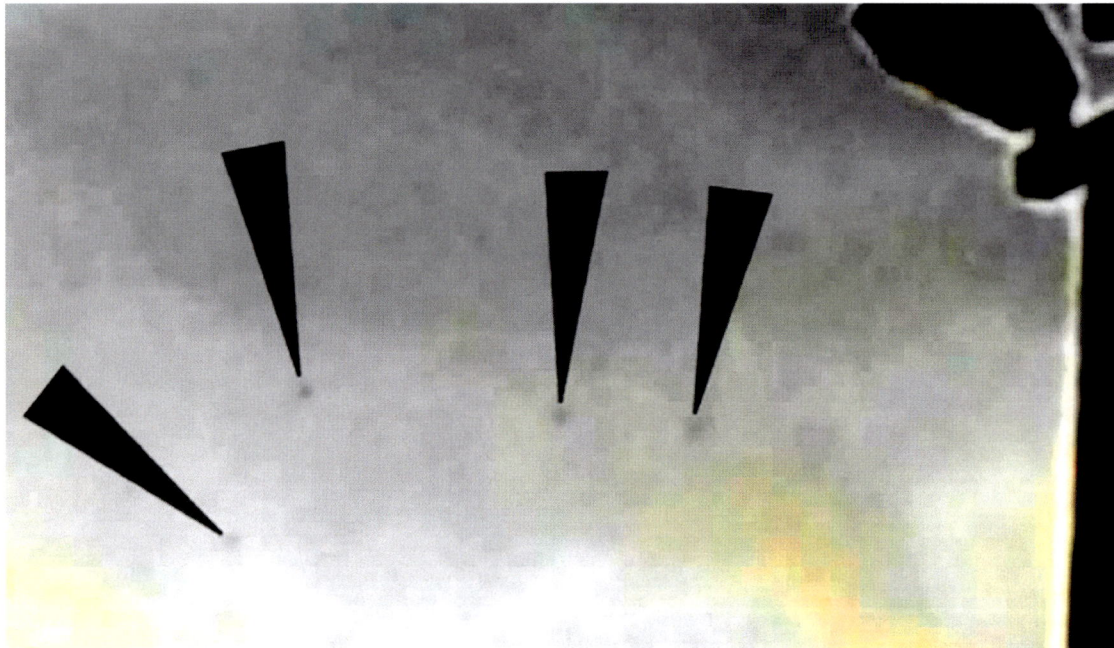

T = 4 close-up & enhanced.

T = 6

T = 7

T = 8

T = 9

T = 10

T = 11

T = 12

T = 13

T = 14

T = 15

T = 16 (arrows with red for clarity)

T = 19

T = 20

T = 23

T = 24

T = 30

APPENDIX 3:
Doing The Math

In this section I go over some of the math used to get the numbers I cite in the body of my book. These are meant as quick approximations. In the sections below I will also often be referring to myself in the third person as the "observer" for the sake of convenience and clarity (so as to keep the writing simple). Anyway, if you like geometry and algebra, you're gonna love this...

ALTITUDE, SIZE & SPEED

Although all 4 of the main pictures are very useful, #1677 is the best starting point for doing measurements especially altitude, because it has very clean lines of sight to the surface below the objects. Actual street surfaces, trees, light posts, etc. are all present and all have easily obtainable dimensions. All can be measured and mapped. The result is the heavily detailed map on the next page...

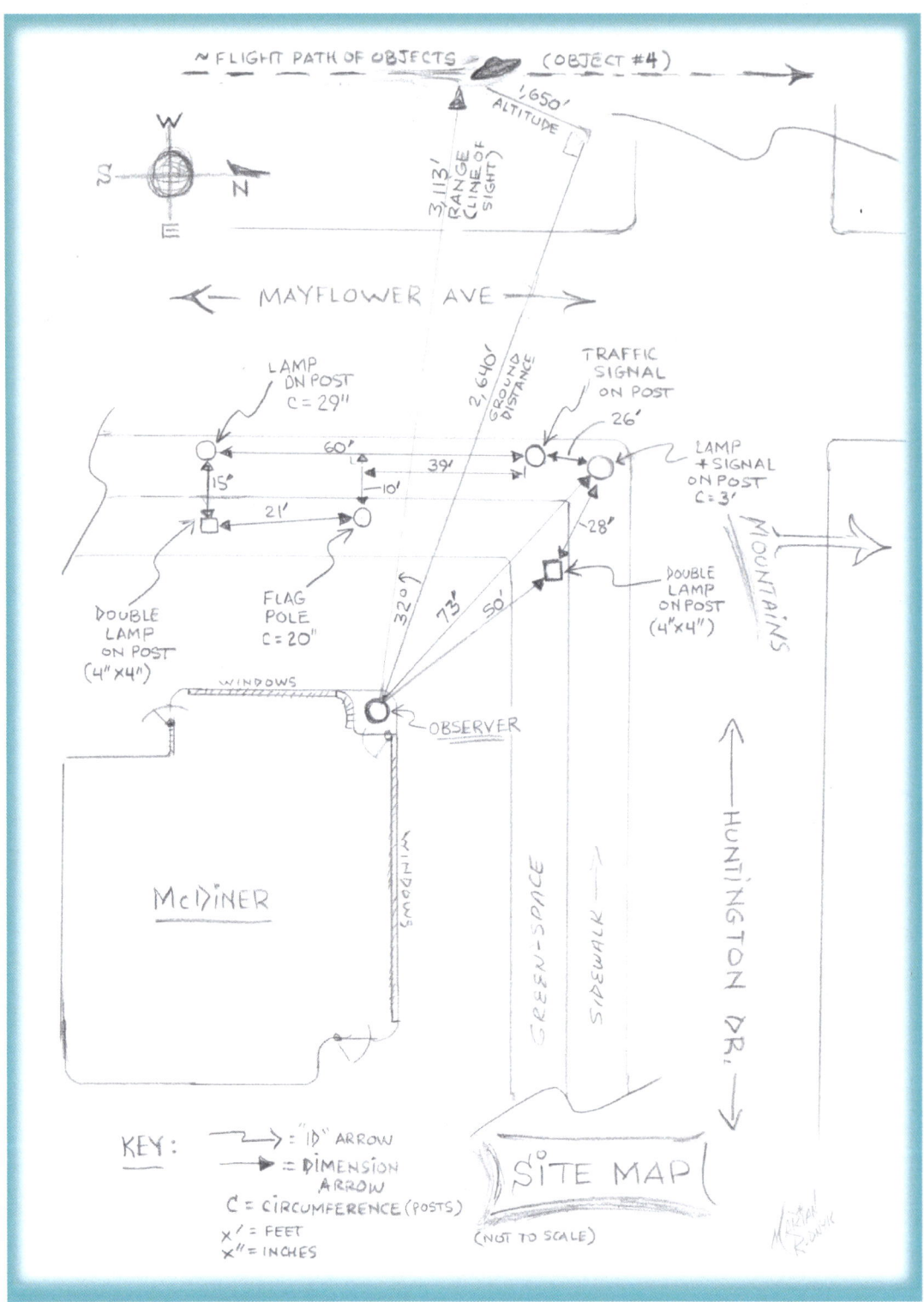

Diagram showing all the measurements I took of the area within the sighting that appears in my imagery. (Detailed sighting area map; ©2017 Marian Rudnyk

As can be seen from the preceding diagram, in the case of this sighting, the pictures and the location provide a very clear 3D position of the objects, confirmable by both the observer, as well as easy accessibility to the site. The math, then, becomes very straightforward.

The figure below provides a look at the geometry involved:

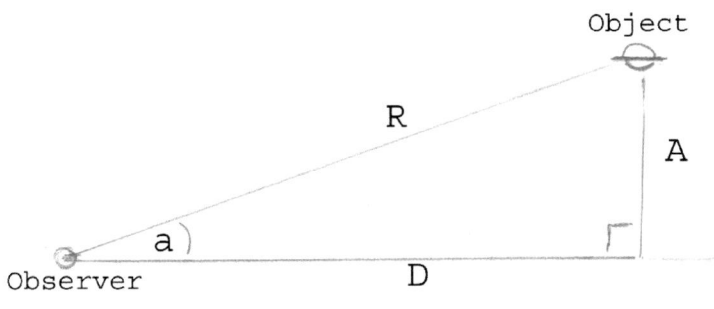

SITE GEOMETRY

Diagrammatic view of the sighting geometry (©2017 Marian Rudnyk), where:

A = Altitude
D = Ground distance from observer to directly below (perpendicular) the object
a = observed angle
R = Range; the direct line-of-sight distance from the observer to the object

This problem (solving for A) requires only the solution of a simple right triangle formula. In this case, the easiest approach is to obtain the viewing angle (a) and the length of the base, (D, for Distance).

Viewing Angle (a):
Object #4 is the best defined object in all the photos, so we will use it. Since frame #1677 has the best ground references (as already described above), we will use this frame for the measurement.

Thus, the viewing angle, measured from the photo, turns out to be 32 degrees. This measurement was confirmed by returning to the site.

Ground Distance (D):
The base of the triangle is actually the location of the objects perpendicularly relative to the ground. Comparison of the images while visiting the site confirms the ground track path, allowing for direct measurement, which was then confirmed by Google maps. The result is a ground distance of about 0.5 miles from the observer to the ground path directly below the objects for frame #1677.

The Calculation: ALTITUDE (A)
So we first start with a simple unit conversion to feet:

1 mile = 5,280 feet
The distance is 0.5miles, so half of 5,280 = 2,640 feet

We need to solve for A (the altitude of the object) so, using our defined terms, we simply use the trigonometric formula and plug in our values:

A = (tangent of a) x D
A = tan(32 degrees) x 2,640 feet = (0.624869) x (2,640 feet) = 1,649.65 ≈ 1,650 feet

Thus we now know that in frame #1677, Object #4 was flying at an approximate altitude of 1,650 feet, or roughly 1/3 mile.

The Calculation: RANGE (aka "line-of-sight" distance)
Using the Pythagorean Theorem we know:

$a^2 + b^2 = c^2$

$c = \sqrt{a^2 + b^2}$

Now substituting our symbols and solving:
$R = \sqrt{A^2 + D^2} = \sqrt{1,650^2 + 2,640^2} = \sqrt{9,692,100} = 3,113$ feet

So we now know the object was about 3,113 feet away, by line of sight

WHAT WE NOW KNOW: RESULTS COMPILED
We now know that in frame #1677, Object #4:

1. is flying at an altitude of 1,650 feet (about 1/3 mi. up)

2. is 3,113 feet away (line of sight range)

3. flying directly over a spot 2,640 feet away from the observer (ground distance); so in other words, you would have to run to a spot 2,640 feet (or ~1/2 mi.) to stand directly below Object #1.

SUMMARIZING:
Looking at the other objects and frames we can summarize that, generally speaking, Objects #1-4 flew roughly between 1/3 mile (up to possibly 1 mile) altitude, at a distance of ~1/2 mile away, and increasing as the sighting progressed.

SIZE

Knowing the altitude it is now easy to take frame #1677, and simply create a conversion scale:

1,650 ft = 741 px or $\frac{1,650 \text{ ft}}{741 \text{ px}}$ = 2.2267 or ~ 2.3 ft/px (the value is rounded to allow for blur)

where: ft = feet
px = pixel

Measuring Object #1 gives a value of 8 pixels (for objects at that distance).

Simply multiplying we get:

~2.3 ft/px x 8 px = ~18.4 ft

So we now know that the Objects are about 18.4 feet in diameter. Although this is a rough estimation, it does appear to fit with what was observed.

As an added exercise to see the +/- effect we can do the following as a check

- If we had measured 7 px (x2.3 ft/px) gives a value of 16.1 ft diameter as an lower range

- If we had measured 9 px (x2.3 ft/px) gives a value of 20.7 ft diameter as an upper range

So we know that, with a range of between 16.1ft – 20.7ft, our answer of 18.4ft is a safe median since +/- 1 pixel does not significantly affect the size. So we can use it for the rest of our calculations.

Measuring the heights gives an average value of roughly 2.6px

Thus 2.6px x 2.3ft/px = ~ 6ft

Comparing these numbers to the other objects yields similar results.

In summary, we can therefore safely conclude that the objects/craft have the following dimensions:

Diameter = ~ 18.4 feet

Height (central) = ~ 6 feet

SPEED

Watching the video reveals that, although the objects move in a generally in a linear overall path, at fairly slow speed. Object-5 is the exception. It is proof that these craft have the ability to move with great bursts of speed, if necessary. It is recommended that future studies establish its speed, though that may prove difficult based on information available. The only thing one can safely say is that Object-5 was really far away, and in the next frame (barely 12 seconds of less later) was significantly closer. Definitely a major burst of speed. However, the mean speed for Objects 1-4, can easily be measured by plotting the positions onto frame #1677, and using Object #1 as a base. This is especially true since the objects travel roughly perpendicular to the view during most of the encounter (and thus minimally impact the pixel scaling factor).

Object-1 travels ~ 1,900 px in ~2 minutes

Thus, using our pixel conversion factor, we get: 1,900 px x 2.3 ft/px = 4,370 ft

Converting to miles we get: 4,370 ft/2 min. x 5,280 ft/mi. x = ~ 0.41 mi. per minute

Now to convert to mph (miles-per-hour) we do:
        0.41 mi/1 min. x 60 min./ 1 hr. = 24.6 mph or, rounding = ~ 25 mph
        - Checking this against the other 3 objects yields similar results.

In summary, Objects # 1-4 all appear to move an average speed of about 25 mph. This is consistent with their appearance in the video and the description by the observer. Their speeds do vary, as the shift formation, but only minimally.

MOVEMENT

A careful review of all the images yields a surprising amount of information about the objects' movements. The figure below shows a plot of the positions of the 4 main objects during the total 1 minute 14 seconds of the photographic portion of the sighting – which includes pictures and the video. Object positions during the video filming are not included because generating them is beyond the scope of this book (however, a glance through *"Appendix – 2: Video Screenshot Gallery"* clearly shows the objects movements are consistent with the pictures). For this figure, a red square roughly defines the area that was covered by most of the video. Viewing the video shows the objects shift position as they try to regain their formation. This is evidenced by their positions in the first picture (represented by the yellow dots), as compared to after the video sequence (the red box area), by the last three photos (as represented by the red, blue and green dots).

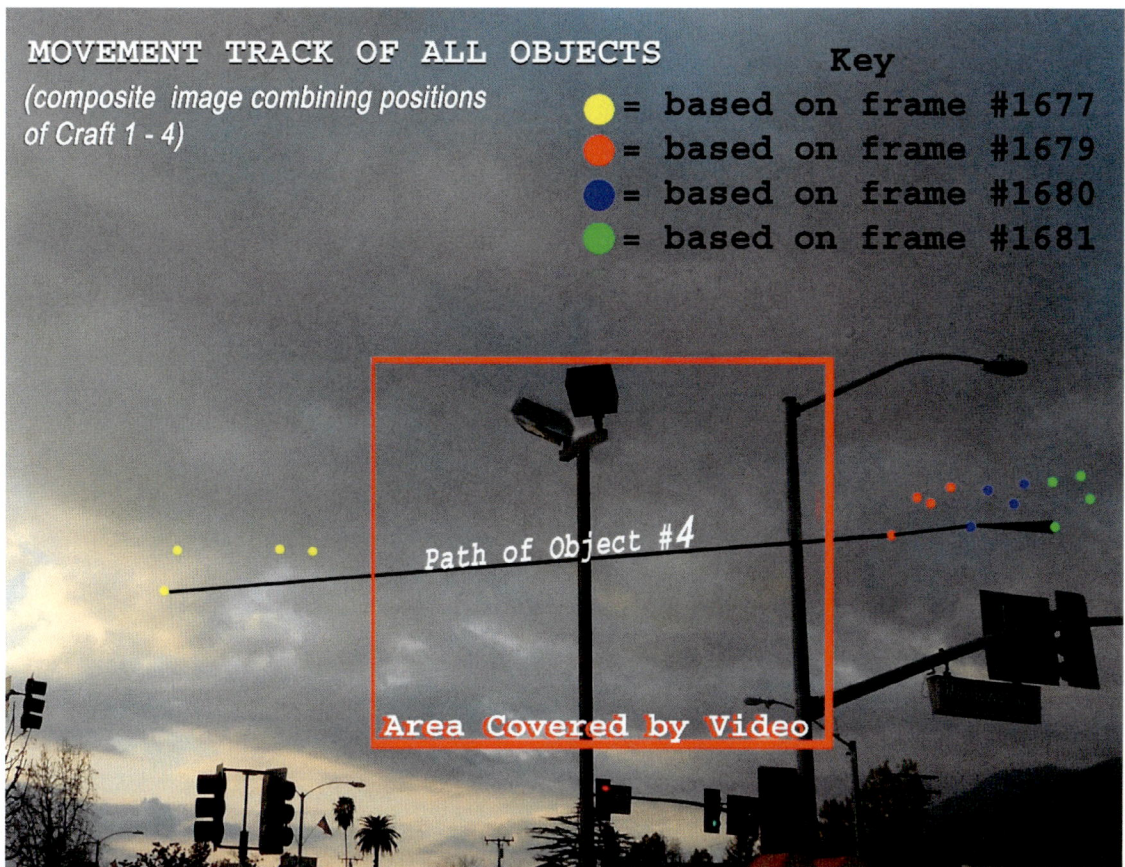

Composite montage image showing movement of the 4 main objects/craft. Combined like this it is easy to see how their formation broke due to the crippled Craft-4 and how they regained formation just before they escaped into the clouds. (Compiled UFO movement graphic ©2017 Marian Rudnyk)

OBJECTS # 1 - 4

The striking thing about the movements of Objects #1-4, is that they were in a diamond formation when initially observed at the start, during the unphotographed portion of this sighting (when I first spotted them from inside the McDiner). As Object #4 began lagging, the formation began spreading out and breaking apart. This is confirmed by the first image at the start of the marked track and as their movement's progress. Objects # 1-3 somehow compensate and "help" Object #4, so that by the end of the sighting, the formation is basically intact, with Object #4 still slightly low, but following and maintaining formation as the objects disappear into the haze around the mountains. This last part is also important, because after the objects were photographed by the observer for the last picture, the observer turned away and glanced inside the restaurant, and when he turned back to take another look at the objects, all 4 were simply gone. Because the clouds by the mountains (towards which they were heading) were still a significant distance away, this directly implies that the objects had a major burst of speed in order to disappear into the cover of clouds and mountains. How long that burst was, and how far they went is not known, but with one of the craft being damaged, I can't imagine they got very far – only far enough away to get to cover.

CONCLUSIONS

From all of the above we can conclude that the objects are disk shaped, about 19 ft. in diameter, and about 6 ft. high in the middle. They can maintain slow speeds, but have the ability of high-speed bursts, or more, whenever they choose. They flew about 1/2 a mile away from the observer and flying at an altitude of about 1/3 of a mile.

APPENDIX 4:
Support Materials

It may seem trivial, but for the sake of completeness, here is a copy of the McDonald's McDiner dining receipt. It supports the sighting's date, time, location and address:

The original McDonald's McDiner (Store #17361) receipt (at left), dated Sunday Jan. 1, 2017 at 14:51. It also shows the address of 480 W. Huntington Dr., Monrovia, CA.
(McDiner receipt photo is ©2017 Marian Rudnyk)

Seen here are my Certificates Of Appreciation from JPL-NASA for my work on the JPL Expo Open House events for 1991. *(Pictures of certificates is ©2019 Marian Rudnyk)*

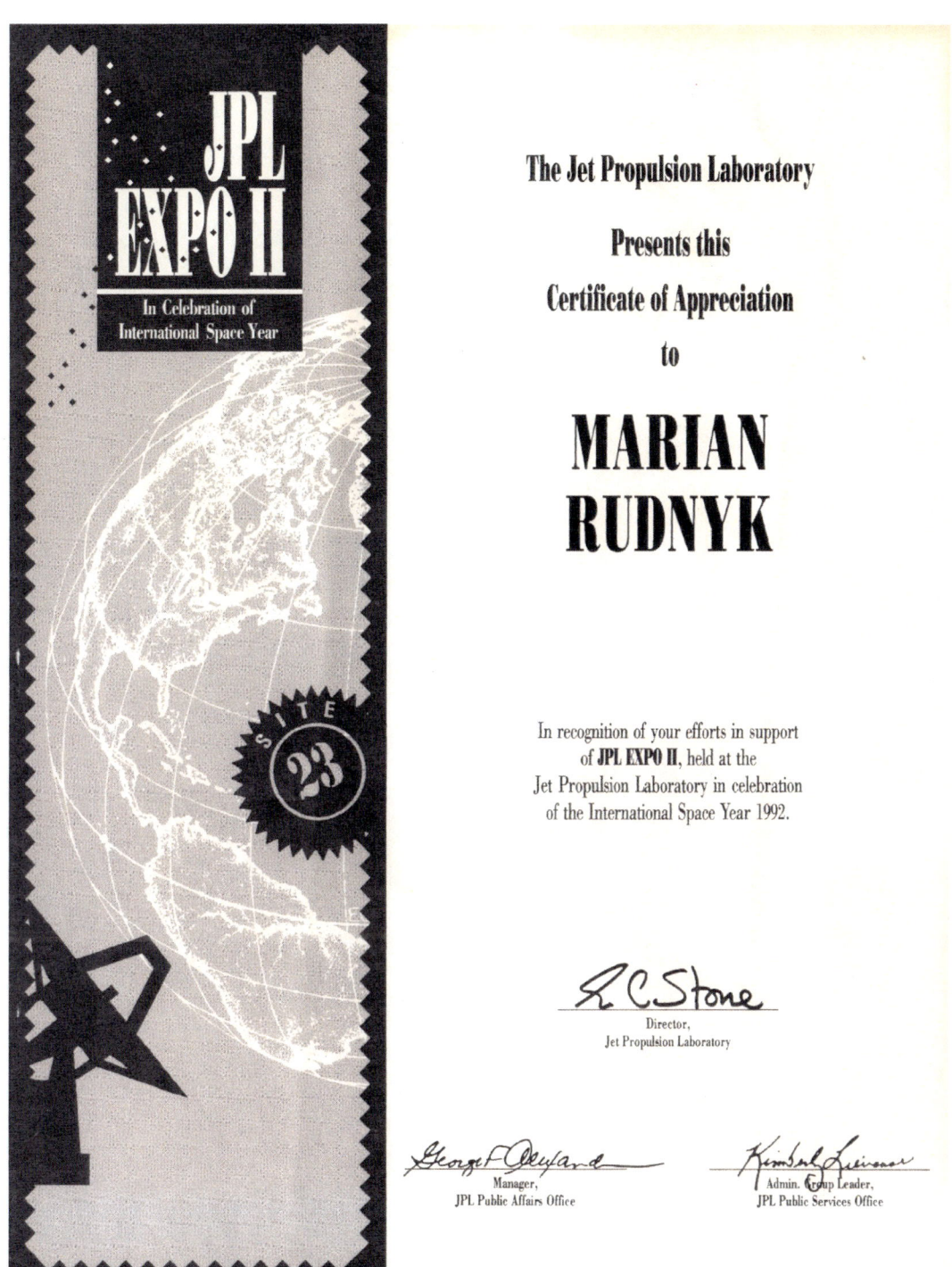

Seen here are my Certificates Of Appreciation from JPL-NASA for my work on the JPL Expo Open House events for 1992. *(Picture of certificates is ©2019 Marian Rudnyk)*

Here is a picture of my Group Achievement Award certificate from NASA for my work on the Neptune Encounter as part of the Voyager 2 Flight Team, dated Sept. 19, 1990. I was part of Multimission Image Processing System Operations. *(Picture of certificate is ©2019 Marian Rudnyk)*

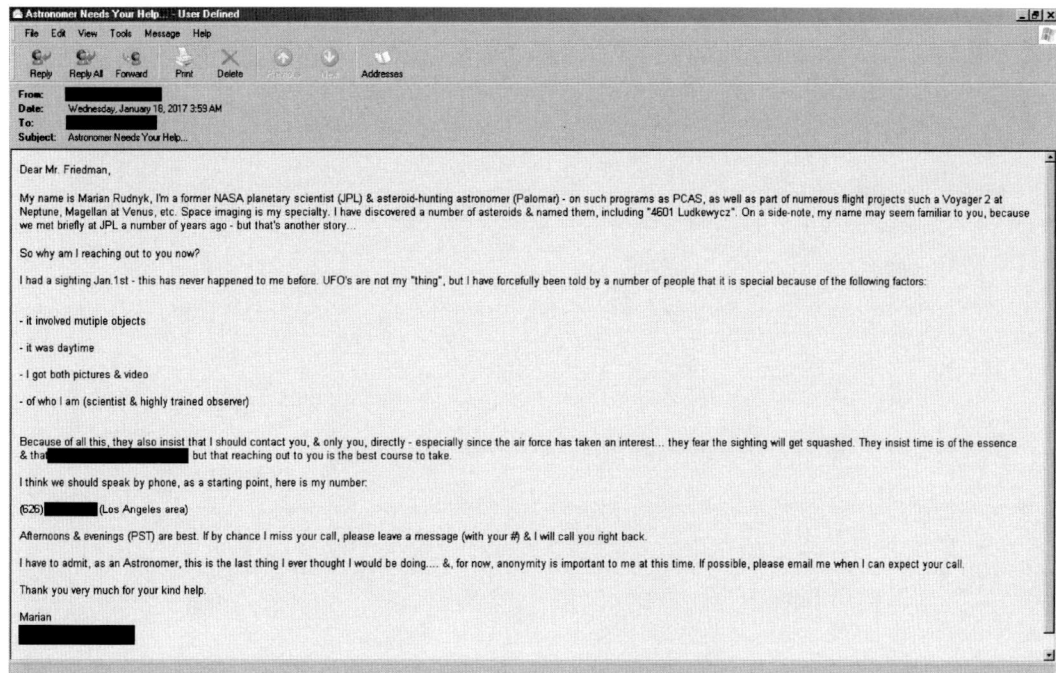

(This image of the email to Stanton Friedman, dated Jan. 18, 2017 is ©2019 Marian Rudnyk)

Above is a picture of my email to Stanton Friedman, it reads:

Dear Mr. Friedman,
My name is Marian Rudnyk, I'm a former NASA planetary scientist (JPL) & asteroid-hunting astronomer (Palomar) - on such programs as PCAS, as well as part of numerous flight projects such a Voyager 2 at Neptune, Magellan at Venus, etc. Space imaging is my specialty. I have discovered a number of asteroids & named them, including "4601 Ludkewycz". On a side-note, my name may seem familiar to you, because we met briefly at JPL a number of years ago - but that's another story...
So why am I reaching out to you now?
I had a sighting Jan.1st - this has never happened to me before. UFOs are not my "thing", but I have forcefully been told by a number of people that it is special because of the following factors:
- it involved mutiple objects
- it was daytime
- I got both pictures & video
- of who I am (scientist & highly trained observer)
Because of all this, they also insist that I should contact you, & only you, directly - especially since the air force has taken an interest... they fear the sighting will get squashed. They insist time is of the essence & that mufon is not the answer now, but that reaching out to you is the best course to take.
I think we should speak by phone, as a starting point, here is my number:
(626) xxx - xxxx (Los Angeles area
Afternoons & evenings (PST) are best. If by chance I miss your call, please leave a message (with your #) & I will call you right back.
I have to admit, as an Astronomer, this is the last thing I ever thought I would be doing.... &, for now, anonymity is important to me at this time. If possible, please email me when I can expect your call.
Thank you very much for your kind help.
Marian
XXXXXX@XXXX.xxx

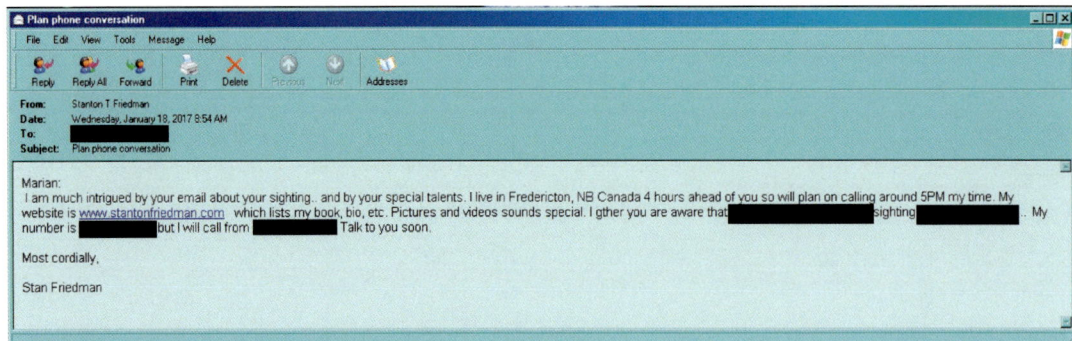

(This image of the email to Stanton Friedman, dated Jan. 18, 2017 is ©2019 Marian Rudnyk)

Above is a picture of my email from Stanton Friedman, it reads:
Marian:
I am much intrigued by your email about your sighting.. and by your special talents. I live in Fredericton, NB Canada 4 hours ahead of you so will plan on calling around 5PM my time. My website is www.stantonfriedman.com which lists my book, bio, etc. Pictures and videos sounds special. I gther you are aware XXXXXXXXXXXX of sighting XXXXX... My number is XXX-XXX-XXXX but I will call from XXX-XXX-XXXX. Talk to you soon.
Most cordially, Stan Friedman

I then reached out to Bob Wood. We had a nice and lengthy conversation and afterwards he sent me a nice quick email note on Sunday, February 05, 2017 3:56pm that said "Good to talk to you today". We had more phone talks, and I tried to send him pictures from my Jan. 1, 2017 sighting but the MIBs found out and stopped it, sparking this reply:

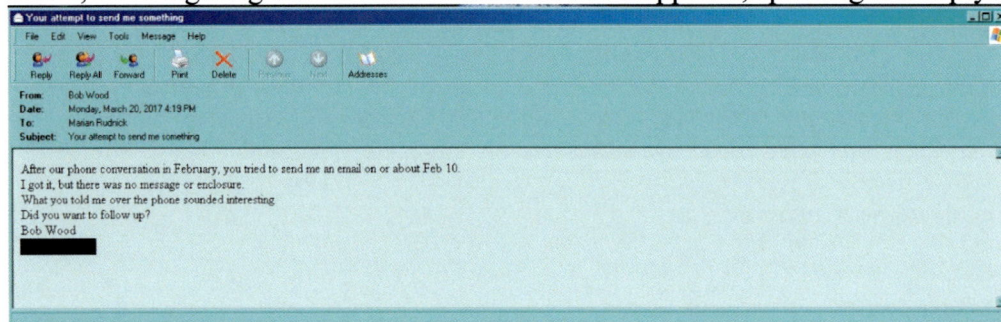

(This image of the email from Bob Wood, dated Jan. 18, 2017 is ©2019 Marian Rudnyk)

Above is a picture of my email from Bob Wood, it reads:
From: Bob Wood
To: Marian Rudnick
Sent: Monday, March 20, 2017 4:19 PM
Subject: Your attempt to send me something
After our phone conversation in February, you tried to send me an email on or about Feb 10. I got it, but there was no message or enclosure.
What you told me over the phone sounded interesting.
Did you want to follow up?
Bob Wood
XXX-XXX-XXXX

We managed to talk more, but it became clear, that for now it was better to wait.

Above is a NASA-JPL sign in log from June 1990 showing Erik Beckjord signing in on June 25, 1990 for the purpose of working with "VO" – which stands for Viking Orbiter imaging data set of Mars. And below is the service request showing him receiving spare Viking Mars maps June 28, 1990. *(Both images are ©2019 Marian Rudnyk)*

Note: personal information of other researchers and scientists has been redacted for privacy and security.

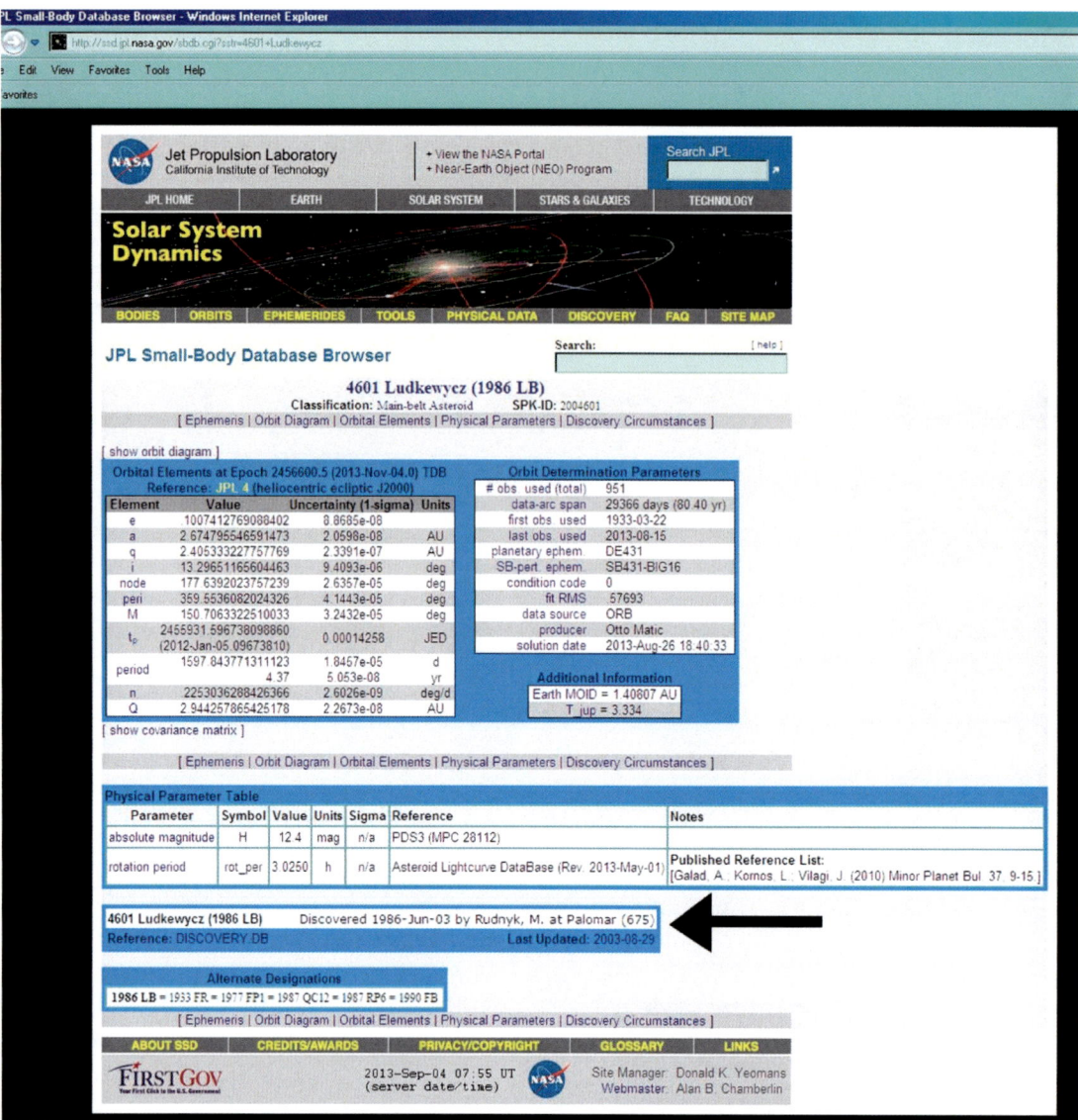

Screenshot of my first asteroid discovery, asteroid 4601 Ludkewycz, on the NASA's JPL Small-Body Database, showing my discovery information as well as orbital elements. (Image ©2017 Marian Rudnyk.)

APPENDIX 5:
Original Sighting Stats

Observer: Marian Rudnyk
former NASA Astronomer & Planetary Scientist
Monrovia, California

Date: Sunday, January 1, 2017.

Sighting: 5 unknown aircraft/craft/UFOs (aka "Objects" – identified by number) exhibiting advanced flight characteristics.

Detail: A total of 4 craft of indeterminate nature and origin, flying in formation, were sighted and were both photographed (4 photos) and a video (one) was recorded. Later examination of the 4 photo frames revealed 1 additional distant craft that was not noticed by the observer (because of its distance and its flight path mostly lined up with a lamp post). All 4 objects exhibited advanced flight characteristics: no flight surfaces, noiseless, ability to accelerate instantaneously, no "visible" or "evident" propulsion system. One craft appeared to be crippled/damaged.

Time:

- Complete event timeline: ~2:45pm to ~4:45pm: about 2 hours total

- Main sighting of 4 main craft: 4:22pm to 4:24pm: about 2 minutes

- Total time range of all photos: 2:47:26 to 4:24:36 = 1hr. 37min. 10sec.

- Time Range of photos of objects: 4:22:58pm – 4:24:12pm = 1 min. 14 sec.

Location: McDonald's McDiner
480 W. Huntington Dr.
Monrovia, CA 91016.
(626) 932-0102

N 34° 08.407 W 118° 00.555
11S E 406949 N 3778152

Weather & Conditions:

Heavy clouds. Partly clearing by late afternoon with many heavy low clouds remaining. Earlier it had been raining intermittently. Temperatures were cold (for California) ~ 62 F (drove by the Citi Bank temp-clock sign on the corner of Myrtle & Palm in Monrovia to get temperature just after the sighting) and confirmed it on home thermometer. Winds were occasionally gusting, light to strong, and generally west to east (as confirmed by pictures, see below). By the time of the sighting, after 4pm, the sun was beginning to set.

Comparison of massive over-sized flag in background, showing gusty winds blowing from west to east. (Multi-frame montage showing flag; ©2017 Marian Rudnyk)

The image montage above, demonstrates 2 important things...

First off, it corroborates the observers reporting of gusting winds. The flag in the background, in the parking lot of the Living Spaces furniture store across (north) the street from the McDiner is a massively over-sized flag. Even so, the winds were strong enough that day to move it considerably as evidenced in these randomly acquired back-to-back frames from the day of the event, and just prior to the sighting.

The other thing these two frames demonstrate is the possible importance of all the pictures shot on the day of the sighting. There is always the possibility that such a frame, seemingly innocuously innocent, may be key to revealing something important. It is for this reason that, not only were all the frames preserved, but are also archived and indexed in the Appendix.

Additionally, the audio in the video supports the often strong gusting nature of the winds. In it you can clearly hear the near constant clanging of the flag-line against the pole of the flagpole at the McDiner. This flagpole is also visible in the photos. Any aircraft flying below the cloud deck would be buffeted by these strong winds.

EPILOGUE

With everything that has happened and continues to happen, even as I write this, I felt compelled to somehow end on something positive and nice. So here are two short, but important, shout outs…

1. HAPPY ANNIVERSARY APOLLO 11!
This year being the 50th anniversary if Apollo 11 landing the first man on the moon I thought it only appropriate to share a nice Apollo memory:

As I a kid who was hyper focused on science and astronomy, one year I felt particularly lucky. Apollo 17, the last U.S. mission to land on the moon, had returned on December 19, 1972. The Command Module was brought back by the recovery ship, the aircraft carrier USS Ticonderoga, to port in Southern California. At the same time NASA had a traveling display that was touring the area that included an actual experimental lifting body vehicle. The traveling NASA show was to be at Sea World at San Diego, and at the last minute NASA had decided to add the actual Apollo 17 Command Module, "*America*", that had flown to the moon, to the show! I was lucky enough to actually see it…

Me as a kid with my family, next to the Apollo 17 Command Module shortly after its return from the moon, at Sea World, San Diego, California. (Pictured left-to-right: my mom, me flanked by my two sisters, and my brother Adrian (right, standing solo). (Photo from personal archive collection taken December 1972; ©2019Marian Rudnyk)

The capsule was strapped to an open flat bed and included as a surprise to those lucky enough to be there at Sea World on that one day – and we were there! In an amazing sign of the times, anyone lucky enough to be there could also actually touch the craft, and pose next to it for pictures.

2. A SPECIAL THANK YOU

Last but certainly not least: with everything that was, and still is happening, it often felt as if I was all alone in all this, but the reality is that I had some very incredible and amazing people backing me up every step of the way. Not only was my brother there for me to help navigate all this madness, but there were two people without whom none of this would even have been possible: Yvonne and Jeanine! You know why: so thank you! Your continued support means the world to me. Additionally, there are Hannes and Rachel who have stood by me, believed in me, and continue to help out when times are toughest – there's a good reason my mom calls Hannes, "Santa Hannes"… thank you! And Rachel, thank you for also encouraging me to "follow my path". There is also one who cannot presently be named. You know who you are, and you know why you cannot be named, and I understand. You have my respect and gratitude. And a quick shout out to Anthony and Josh: your enthusiasm is infectious! Never forget the "red dot"! Thanks dudes! Thank you Marko. Big thanks to FanGirl for the tech help – you're the best! Thanx Taras – hang in there! Romtsja – thanks for finally believing me! Bob – thank you for being such a good friend, even in hard times. Thank you Stanton Friedman and Bob Wood for your support and giving me perspective when I needed it most. And last but undoubtedly not the least, my mom, who has been there with me from the very start, and every step of the way, and lovingly and with unending support, endured all this craziness. Thanks mom. Thanks to all of you!

…And finally, a friendly reminder to make sure to pick up a copy of the much anticipated sequel to this book:

"INTERSECT 2: The Game Changers"

- coming soon during the winter of 2019!

Manufactured by Amazon.ca
Bolton, ON